Punctuation Decoded

by Amanda A. Knight

Photos by Rachel E. Knight

Printed in the United States of America.

ISBN: 1-60250-036-3
978-1-60250-036-5

59 Damonte Ranch Parkway, Suite B284 • Reno, NV 89521 • (800) 970-1883

www.benttreepress.com

Address all correspondence and order information to the above address.

With special thanks to David,
my greatest fan,
who has always encouraged and believed in me....

Introduction

Over the last twenty years, I have developed a method to teach students to compose better, more-complex sentences and to punctuate those sentences correctly. As an added benefit, by studying my method, students also learn not to make major sentence construction errors. Over the years, I determined that there is no way to teach punctuation effectively if sentence structure is not taught at the same time. **Sentences are punctuated according to their clause structure.** That is a basic fact, and students cannot be expected to learn punctuation if they do not learn to identify the structure of sentences. Over the years, I also discovered that students will not learn to identify clause structure just by looking at a sentence. In frustration, I remembered the "old school" method of diagramming sentences, and although I do not teach diagramming, I do have students physically mark the clauses in a sentence. Through this, they become visually engaged with the sentence, and by physically identifying clauses, they study the punctuation that the clauses require. My method works; I've used it for many years at the high school and college level. Students who previously found commas to be arbitrary and semicolons to be otherworldly have discovered that this method makes sense to them. I also endorse abandoning English-teacher jargon, words that many students find very intimidating, terms like "subordinating conjunctions" or "conjunctive adverbs." I urge the usage of less-daunting expressions, like "dependent words" or "transitional expressions." The concept is still taught, but from a fresh perspective, which many students find more appealing. I hope that you can benefit from my method. It is actually a very simple concept: once students can identify clause structure and apply a few basic rules of punctuation, they are capable of creating complex sentences to express themselves confidently for the rest of their lives.

Please let me know if my method works for you, too.

Wishing you many "ah-ha moments,"
Amanda A. Knight
Associate Professor of English
Andrew College
Cuthbert, Georgia

Contents

Truth #1: All sentences must have a subject and a verb.

Chapter One: Identifying Simple Subjects and Simple Predicates (Verbs)....1
Justification....1
Clarifications....1
The Handy, Dandy List....2
Exercise 1....4
Exercise 2....5
Exercise 3....7
Progress Check Quiz....9

Truth #2: All sentences must make a complete thought.

Chapter Two: Identifying Independent and Dependent Clauses....11
Justification....11
Clarifications....11
Exercise 1....13
Exercise 2....14
Exercise 3....16
Exercise 4....18
Progress Check Quiz....21
Exercise 5....23
Exercise 6....25

Chapter Three: Combining Clauses....29
Justification....29
Clarifications....29
Exercise 1....30
Exercise 2....32
Exercise 3....32
Exercise 4....34
Exercise 5....35
Exercise 6....36
Progress Check Quiz....39

Chapter Four: Sentence Errors....41
Justification....41
Clarifications....41
Exercise 1....43

Exercise 2 44
Exercise 3 45
Exercise 4 47
Exercise 5 49
Progress Check Quiz 51

Chapter Five: Commas 53
Justification 53
Clarifications 53
Exercise 1 55
Exercise 2 58
Exercise 3 60
Exercise 4 62
Exercise 5 64
Progress Check Quiz 67

Chapter Six: Semicolons and Colons, Parentheses and Dashes 69
Justification 69
Clarifications 69
Exercise 1 71
Exercise 2 72
Exercise 3 74
Exercise 4 76
Progress Check Quiz 79

Chapter Seven: Quotations and Italics 81
Justification 81
Clarifications 81
Exercise 1 84
Exercise 2 84
Exercise 3 85
Exercise 4 85
Exercise 5 86
Progress Check Quiz 87

Chapter Eight: Apostrophes 89
Justification 89
Clarifications 89
Exercise 1 90
Exercise 2 91
Exercise 3 92
Progress Check Quiz 95

Chapter Nine: Capitalization 97
Justification 97
Clarifications 97
Exercise 1 98
Exercise 2 100
Exercise 3 101
Progress Check Quiz 103

CHAPTER *one*

Identifying Simple Subjects and Simple Predicates (Verbs)

TRUTH #1: ALL SENTENCES MUST HAVE A SUBJECT AND A VERB.

JUSTIFICATION: WHY DO I NEED TO KNOW THIS?

Although this may seem like a very elementary skill, being able to identify the simple subject and simple predicate (verb) of a sentence is essential to creating well-constructed sentences, to eliminating sentence errors (run-ons, comma splices, etc.), to ensuring subject/verb agreement within sentences, and to mastering other elements of the sentence. Therefore, please give this chapter the study that it deserves—it will serve you well later. (I guarantee it!)

CLARIFICATIONS

You must be able to identify the simple subject of a sentence. That is, you should identify just the subject (usually just one word), not the modifiers that go along with it.

Example: The gross, smelly garbage has been sitting there all day.

The simple subject is *garbage*; omit *the gross, smelly.*

Also, you must be able to identify the simple predicate (verb) of a sentence. You will need to identify the entire verb (which may include helping verbs), but omit any modifying adverbs. From the above example, you would identify the verb as *has been sitting*. You would omit *there*—it tells where, so it is an adverb. Be on the lookout for adjectives that are very tempting to include as part of the verb. These adjectives appear after linking verbs and describe the subject.

Example: The garbage is smelly.

The verb for the sentence is *is*. The word *smelly* describes; it is not a verb. (To check yourself, ask "Is this an action that is done, or does this word describe the subject?")

I have composed a Handy, Dandy List of words and suggestions that should prove helpful to you. Please memorize the word lists and know the helpful tips for identifying simple subjects and predicates.

CHAPTER 1: THE HANDY, DANDY LIST

(Helpful words and suggestions for identifying simple subjects and predicates)

1) **Identify the verb of the sentence first!** (It's easier that way!) Be sure to include any helping verbs.

   Here's a list:

Common Helping Verbs

am	did	might have been
are	do	must have
be	does	must have been
being	had	was
been	has	were
can	has (had) been	will (shall) be
can (may) be	have	will (shall) have
can (may) have	is	will (shall) have been
could (would, should) be	may	
could (would, should) have	might have	

Also, know that the parts of a verb phrase may be separated from one another by other words:

Ex: Did you empty the trash cans?
I can hardly breathe on account of the odor.
(*You* is the subject of the first example; *hardly* is an adverb.)

2) ***To* + verb is not the verb of the sentence!** (It's an infinitive.)
   Check the word before the verb. If it is preceded by *to*, it is not the verb.

Ex: My husband did not want to empty the trash can.
(*Did want* is the verb of the sentence. *Not* is an adverb, and *to empty* is an infinitive.)

3) **Identify the simple subject after you've identified the verb**. (Ask yourself who or what performed the verb.)

Ex: The busboy took out the trash at the restaurant.
(Who or what took? *Busboy*—that's the simple subject of the sentence.)

4) **The simple subject is never found in a prepositional phrase**. Prepositional phrases always act as modifiers and only provide "extra stuff." (It is often tempting to identify the object of the preposition as the subject—don't fall for it!) Draw a line through prepositional phrases to eliminate them.

Ex: The stench of the foul garbage cans at the restaurant door was disgusting and sickening.

(Eliminate *of the foul garbage cans* and *at the restaurant door* as choices for the simple subject; they are prepositional phrases, so the simple subject cannot be in them. Find the verb first—*was*. Remember that *disgusting* and *sickening* describe; they are adjectives. Now ask "Who or what *was*?" The simple subject must be *stench*.)

Commonly Used Prepositions

about	between	over
above	beyond	past
across	but (meaning except)	since
after	by	through
against	concerning	throughout
along	down	to
amid	during	toward
among	except	under
around	for	underneath
at	from	until
before	in	unto
behind	into	up
below	like	upon
beneath	of	with
beside	off	within
besides	on	without

(Sometimes a group of words may act as a preposition: *on account of, in spite of, along with, together with, because of, etc.)*

5) ***There* or *here* will not be the simple subject of the sentence.** Sentences that begin with the words *there* or *here* will have the subject located after the verb.

Ex: There were many rotten fish heads in the garbage.

(Eliminate *in the garbage*—prepositional phrase—and *many rotten fish*—adjectives. From what is left *There were heads,* the verb is *were*. Who or what *were*? Heads. *Heads* is the simple subject.)

6) **Watch out for the understood *you* subject!** (as in that very sentence) Sentences with an understood *you* as the subject give directions or orders. *You* is the subject, the person to whom the order is addressed or spoken.

Ex: Take out the stinking trash! (Who or what *take*? *You* take.)

7) **Questions are easier to deal with as statements.** Make questions statements; then find the verb and then the subject.

Ex: Do you smell that?
You do smell that. (Who or what *do smell*? *You.*)

8) **Know these conjunctions**—There can be more than one subject and verb in a sentence.

Ex: The stench of the garbage and the appearance of the restaurant made customers nauseous and turned them away.

Eliminate *of the garbage* and *of the restaurant*—they are prepositional phrases. Find the verb—*made* and *turned.* Who or what *made* and *turned*? *Stench* and *appearance*—those are the simple subjects. (*Nauseous* describes and *away* tells where—it's an adverb.)

Coordinating Conjunctions

and but or nor for yet so

Now that you have these tips (and I've sickened you with gross images), apply these steps as you complete your practices:

1) **Read the sentence (and understand it!)**
2) **Omit prepositional phrases (Neither the subject nor the verb will be found there.)**
3) **Identify the verb (Watch out for adjectives, adverbs, and *to* + verb)**
4) **Ask "Who or what performed the verb?"**
5) **Identify the simple subject (Look out for prepositional phrases, *there/here*, understood *you*, questions, adjectives, etc.)**
6) **Proofread your choices (Did you check the entire sentence? Are there conjunctions?)**

(Refer to this page as you do your practice exercises and learn these tips!)

CHAPTER 1: EXERCISE 1

First, read and understand the sentence. Then, eliminate any prepositional phrases by drawing a line through them. Next, identify the verb(s) by underlining them twice, and identify the subject(s) by underlining them once.

1. Jan wanted a cell phone for Christmas, but her parents could not afford one.
2. At the beginning of the new year, Jan got a job and worked at a restaurant.
3. For six months, she saved all of her pay, and her parents were proud of her work ethic.
4. Jan wanted to get a small, slim, sleek phone, but the older, larger model was less expensive.
5. From her new perspective as a money-earner, Jan decided to get the less- expensive phone.
6. She would have more money to spend on her cell phone plan.
7. Because of her smart thinking, Jan has a new phone and can talk for hours and hours.
8. She is not going to switch plans but is looking at other options.
9. For her birthday, Jan's parents bought her some of the cool accessories for her phone.

10. Now Jan is having great conversations with her friends and is keeping her job at the restaurant.
11. She has learned an important lesson: she can work and earn money to buy fun things.
12. On her next payday, Jan will make a payment toward a Wii, the newest Nintendo gaming system.
13. Jan must earn enough money to buy the game system, and she must try to get a better TV for her room.
14. By requiring the player to move, this new game system may keep Jan more physically fit.
15. Her job at the restaurant does not pay very much; she only works about twelve hours each week.
16. Jan is happy to work; at least she has a job.
17. Her customers always tip her generously, and she makes more in tips than in hourly wages.
18. Jan must have been very nice to one customer; an older lady left Jan a $20 tip once.
19. From that experience, Jan has learned to be pleasant to all of her customers.
20. Working as a waitress can be difficult, but Jan likes to make her own money.

CHAPTER 1: EXERCISE 2

First, read and understand the sentence. Then, eliminate any prepositional phrases by drawing a line through them. Next, identify the verb(s) by underlining them twice, and identify the subject(s) by underlining them once.

1. Joe and Anne are expecting their first baby, and they are a little scared.
2. Like most newlyweds, they were not trying to have a child yet, but sometimes things just happen.
3. Anne has noticed some very strange changes in her body, and Joe is not sure what to expect in the delivery room.

4. At first, Anne noticed that she was tired all of the time, and she slept a lot.
5. Joe was thrilled to hear the baby's heartbeat, and he was relieved to be told that the baby was healthy.
6. Now, Anne's previously slim figure has been replaced with a very large stomach, and Joe is a bundle of nerves.
7. They attended childbirth classes, but Joe couldn't watch the films.
8. Everyone has been excitedly awaiting the birth, but the baby seems to have her own schedule.
9. Anne is miserable and feels awkward and fat; Joe feels excited and wants the baby to get here.
10. They have decided to call her Elizabeth Anne, and Joe has prepared the baby's room with all kinds of gadgets and toys.
11. Anne finally started to have labor pains, and Joe hustled her to the hospital.
12. However, after a full day of observation, Anne and Joe were sent back home; nothing seemed to be happening.
13. Anne was glad to leave; she needed a pedicure, a shower, and a good night's sleep.
14. Joe was about to go crazy; the waiting was becoming too much.
15. Anne went to a salon, got a pedicure, came home, took a shower, and went to bed.
16. Joe puttered around the house; he cut the grass and weeded the garden.
17. Finally, Elizabeth Anne decided that she was ready to make her appearance, and Anne and Joe went back to the hospital.
18. After a lengthy labor, Anne delivered a beautiful baby girl, and Joe finally could hold his daughter in his arms.
19. Joe was a bit pale, but he held up fine through the delivery.

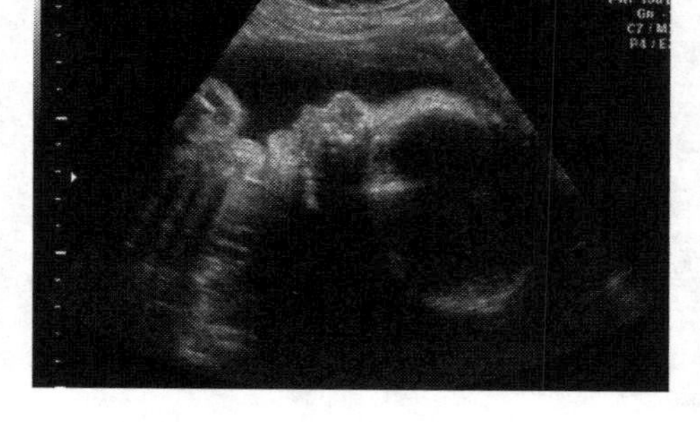

20. After the baby was cleaned up, she was taken out to meet all of her adoring family, and Joe beamed with pride.

CHAPTER 1: EXERCISE 3

First, read and understand the sentence. Then, eliminate any prepositional phrases by drawing a line through them. Next, identify the verb(s) by underlining them twice, and identify the subject(s) by underlining them once.

1. Gardening is in my blood; my mother gardens, and my grandmother gardened.
2. I have tried to keep this compulsion under control, but every spring I have the almost uncontrollable urge to plant something.
3. After the last freeze every year, the garden shops and home improvement stores put out flats of colorful annuals, and I am drawn to them.
4. I love to look at the tiny plants and to envision how they will look in my flower beds at home.
5. This year I did not buy any new plants; our section of the country is experiencing a drought, and it is hard to keep anything alive.
6. I still have potted plants on my porch, and the daylilies have come back like they do every year.
7. Daylilies are hardy plants and take very little care; I love them for that and for their beautiful blooms.
8. Many people in town tell me that my garden is lovely, and I am glad that they find pleasure in my flowers, too.
9. Having an attractive yard can add to the value of my property, too.
10. I have noticed that many people do not take pride in their yards anymore, and they seem not to care about gardening.
11. Growing flowers is not difficult; they just require good soil and the right amount of light, water, and fertilizer.
12. Plants will indicate if they are unhappy; they will drop leaves, turn yellow, or fail to bloom.
13. A gardener just needs to notice these signs and to correct them quickly, and the plant will usually recover.

14. I enjoy spending time with my flowers; every day I remove some dead blooms, pull a few weeds, and water a few plants.
15. As I do these things, I feel a sense of peace, and I can wonder at the beauty of nature.
16. Many times I see insects as they work among the plants; ants scurry about, butterflies float by, and bees buzz in the blooms.
17. For a while, a toad lived at the base of some of my plants, and it would frighten me by hopping out suddenly.
18. I give my plants water, and they give me therapy.
19. I must have been born with a green thumb.
20. I hope that my daughter will inherit my love of gardening; it should be in her blood, too.

Name ______________________________ Date__________

CHAPTER 1: PROGRESS CHECK QUIZ

First, read and understand the sentence. Then, eliminate any prepositional phrases by drawing a line through them. Next, identify the verb(s) by underlining them twice, and identify the subject(s) by underlining them once.

1. Video games are a very popular form of entertainment for college students.
2. Many students spend hours playing a game and become very skilled.
3. Some dormitories with lots of participants sponsor tournaments between players, and trophies or prizes are awarded.
4. Players claim that playing a video game relaxes them.
5. Some gamers develop lightening-fast reflexes and have great hand-eye coordination.
6. Some others play only on the weekends when they have no homework to do.
7. Playing too often can take time away from studies, and grades can suffer.
8. Wise students will not use study time to play video games, and they will limit their play.
9. Foolish students will ignore their need to study and will play video games day and night.
10. Some gamers will fail classes and will drop out of college.

There is a saying: "Those who play don't stay."

CHAPTER *two*

Identifying Independent and Dependent Clauses

TRUTH #1: ALL SENTENCES MUST HAVE A SUBJECT AND A VERB.
TRUTH #2: ALL SENTENCES MUST MAKE A COMPLETE THOUGHT.

JUSTIFICATION: WHY DO I NEED TO KNOW THIS?

Being able to identify the clause structure of a sentence will enable you to punctuate the sentence correctly and create sophisticated, complex, impressive sentences. This ability will eliminate your making serious sentence errors or will enable you to identify and correct such errors (thereby saving you serious points on your written work!)

CLARIFICATIONS

Every sentence must have a subject and a verb, but every group of words having a subject and a verb may not be a sentence. (Read that again and be sure that you understand it.)

Check out this group of words:

Because Tom wanted to bake a cake for his mother's surprise birthday party, which was planned for the following Saturday afternoon at the Lion's Club.

> Is it a sentence? It does have a subject and verb (two, actually–*Tom wanted, which was planned.*) However, it is not a sentence. What do you think? No, it's not a run-on sentence, either. It is a sentence fragment. It is an incomplete thought. After all, the reader is never told "*because Tom wanted to bake a cake*...what?" One word makes this incomplete....can you figure out what it is? The word *because* makes this dependent on another thought to finish it. Such groups of words, which have subjects and verbs but are still incomplete, are called **dependent clauses.** They are fairly easy to identify if you are familiar with two groups of words: relative pronouns and subordinating conjunctions, but we will just call them **dependent words. Memorize this list** (or you will not be able to master sentence errors and punctuation!)

Relative pronouns –These are often used as subjects of dependent clauses.

who whom whose which that what

i.e., The party was a success, *which* surprised Tom.

Common Subordinating Conjunctions* -These often introduce dependent clauses.

after	**as though**	**(ever) since**	**when**
although	**because**	**so that**	**whenever**
as	**before**	**than (not *then*)**	**where**
as if	**if**	**though**	**wherever**
as long as	**in order that**	**unless**	**while**
as soon as	**provided that**	**until**	

i.e., Tom's mother cried *when* she saw all the people.

*Sometimes these words may be used as prepositions: *after the party, until the end.* Look for a subject and verb after them, i.e., *after he sang, until she ate,* to see if they are being used to introduce dependent clauses.

**That* is often used as a pronoun, ex: *That* is my favorite, or as an adjective, ex: *That* food is my favorite.

Study the dependent clauses in the following sentences:

Eric was studying air conditioning repair, *which* was a very lucrative field.

Daniel studied computer programming *because* he wanted to have flexibility in his choice of profession.

When he had difficulty in his schematics course, Eric went to see his friend Steve, *who* was a professional.

Though he had little experience, Daniel applied to IBM, *where* he was told to come back in a few years.

Notice that the underlined group of words begins with *one of those* ***dependent words*** and has a subject and a verb–they are all dependent clauses. Circle the dependent words used. Notice also the commas that occur before or after many of these clauses. They are either there (or not) because of these dependent clauses. The parts that are not underlined are ***independent clauses***–they contain subjects and verbs but are not introduced by one of those dependent words. Notice that these non-underlined parts could be sentences that make complete thoughts by themselves.

Clauses can be joined in a sentence by using coordinating conjunctions. Know these and check carefully to see what they are joining.

Coordinating conjunctions

and but or nor for yet so

Daniel applied to many different companies *after* he was rejected by IBM, **but** he was disappointed *that* IBM wouldn't have him.

When he aced his schematics course, Eric called Steve, **and** the two of them went to celebrate at Fred's Oyster Bar, *where* they drank beer and ate oysters.

Notice the commas that are used in all of these examples.... The rules that apply to their use are easy to learn ***if you can identify clauses!***

CHAPTER 2: EXERCISE 1

Identify each as an independent clause (**I**) or a dependent clause (**D**).
If it is a dependent clause, what word makes it dependent? Circle it.

_____ 1. If I could wear size two clothes

_____ 2. I would become a clothing model

_____ 3. Models earn good salaries

_____ 4. When they go on location

_____ 5. That they often do their own makeup

_____ 6. Which saves the photographer time and money

_____ 7. Plus-size modeling is becoming popular

_____ 8. Although posing can be difficult

_____ 9. Most locations are very exotic

_____ 10. As long as teenaged girls emulate models

_____ 11. While young girls are very impressionable

_____ 12. If models care about their public image

_____ 13. Being healthy should be a top priority

_____ 14. Because large-sized models are in greater demand

_____ 15. Larger women have greater opportunities

_____ 16. When young girls see larger models

_____ 17. Wherever they go

_____ 18. Until the photo is developed

_____ 19. After the agency is notified

_____ 20. Modeling can be a short-lived, stressful career

CHAPTER 2: EXERCISE 2

Follow these steps for each sentence: 1) Read the sentence, 2) underline the verb(s) twice and the subject(s) once, 3) find the dependent words, which introduce dependent clauses, 4) put the dependent clauses in parentheses, and 5) underline the independent clauses.

Example: (Ever since he was a little boy,) Fred has cooked fabulous meals, and he has lots of fans to prove it.

1. Because company was coming for dinner, Fred prepared a special treat of Oriental food.
2. After Fred boiled chicken for stock, he removed the meat and brought the broth to a boil.
3. Before he added shredded vegetables, Fred added a dash of soy sauce and ginger for seasoning.
4. When the mixture came to a boil, Fred turned off the heat and added a scrambled, raw egg.
5. Whenever he entertains with friends, Fred likes to show off his cooking skills.
6. His guests always are provided with chopsticks, which add to the fun.
7. Guests always giggle when they drop food.
8. Fred doesn't mind cleaning up after his guests leave.
9. Because he is such a good customer, Fred gets a discount at the local Oriental market.
10. He gets many good recipe ideas from a lady inside who gives him authentic recipes from her homeland.
11. People are always glad to be invited to Fred's house for dinner, which makes Fred very happy.
12. He thinks that it is a great compliment to be a popular host.
13. When he goes to other peoples' houses for dinner, Fred is often asked to help cook.

14. Fred thought about going to culinary school when he graduated from college.

15. Because he cooks so often, he invested in a very expensive stove.

16. Fred only uses the finest ingredients, which makes a great difference in the taste of his food.

17. Fred's girlfriend can't cook, although she does decorate the table nicely.

18. If they get married, Fred and his girlfriend could open their own restaurant some day.

19. Fred thinks that he would like to be a professional chef one day.

20. Whenever I am in Cleveland, I always stop by to see Fred and his girlfriend.

EXERCISE 2 REVIEW

Now that you have marked the clause structure of all of the sentences, notice sentences 1-5. They all have something in common. What? Do you notice the commas used in all of them after the introductory dependent clauses? Those all illustrate a basic comma rule: **a sentence that begins with a dependent clause should have a comma after the introductory dependent clause.** Do you see any other sentences that show introductory dependent clauses followed by commas? Write the numbers here (there are five others): ___ ___ ___ ___ ___.

This rule looks like this when you have marked the clauses:

(dependent) , ___independent___ introductory dependent clause followed by a comma

Now look at sentences 6 and 7. These show sentences with independent clauses coming first, followed by dependent clauses. However, one has a comma, and the other does not. Here is another comma rule governed by clause structure: **when a dependent clause follows an independent clause, a comma is used to separate them if the dependent clause provides unnecessary, extra information; "extra stuff,"** we'll call it. Look at sentence 6; the dependent clause gives extra information about using chopsticks: they add to the fun.

This rule looks like this when you have marked the clauses:

___independent___ , (dependent) dependent clause giving extra stuff (= comma)

However, sentence 7 tells under what conditions the guests giggle: they are not constantly giggling; they giggle only when they drop food. This is necessary information that reveals under what conditions the guests giggle. Here is another comma rule governed by clause structure: **when a dependent clause follows an independent clause, a comma is NOT used to separate them if the dependent clause provides necessary, needed information.**

This rule looks like this when you have marked the clauses:

independent (dependent)   dependent clause giving necessary information (NO comma)

Can you find examples of each of those rules elsewhere in the exercise? The more that you examine these rules, the easier they will be to apply. Let's have more practice identifying clauses and studying the punctuation they require!

CHAPTER 2: EXERCISE 3

Follow these steps for each sentence: 1) Read the sentence, 2) underline the verb(s) twice and the subject(s) once, 3) find the dependent words, which introduce dependent clauses, 4) put the dependent clauses in parentheses, and 5) underline the independent clauses.

Example:
Computers, (which are necessary equipment for college students,) have made completing assignments much easier.

1. Most students who can use computers well will have an easier time creating college-level work.
2. Students who do not know how to operate computers will struggle to submit assignments that the professors require.
3. Headers and work-cited pages, which require special knowledge to create, can give students fits!
4. Many instructors ask students to submit their work as attachments, which can provide many advantages for the instructor and the student.
5. If assignments are sent as attachments, instructors know exactly when the assignments were submitted, and they have electronic copies of documents.
6. Students can be assured that their papers are not late and that they cannot be lost.
7. After instructors receive the papers, they may have to print all of the documents, which can be time-consuming and costly.
8. If a student does not create the document using the correct program, the instructor may not be able to open the attachment.
9. Because they have electronic versions, instructors can keep files of previously submitted papers, which can discourage plagiarism.

10. After they complete a few assignments using computers, students feel more confident that they can operate them well.
11. Students quickly learn to save their work periodically so that they will not lose it if the power goes out or something.
12. Some students can write a good paper, but they cannot type well and make many typographical errors.
13. Because some students do not fully develop their ideas, their papers may be very short, and they try to lengthen them by using large font sizes, which take up more space on a page.
14. Other students try to lengthen their papers by triple spacing their lines, or they may make the margins narrow.
15. Most professors will quickly recognize these lame attempts to lengthen weak papers, and they usually deduct points from them.
16. When students use a computer that is available for anyone to use, they must save their work to a removable device because documents are often deleted or stolen.
17. Floppy disks, which save relatively small amounts of information, are quickly being phased out.
18. Every college graduate should be comfortable using computers, for those are the tools that will be used in any field that they enter.

EXERCISE 3 REVIEW

- Why is there a comma after the word *versions* in sentence #9?

- What other sentences illustrate this same rule? (There are six others.) ___ ___ ___ ___ ___ ___

- Do you see any dependent clauses that have two **commas** around them, in front and behind? There are two sentences in which the dependent clause is interrupting the sentence with **extra stuff**, unnecessary information. Can you find them? ___ ___

- There are several sentences in which the dependent clause gives **extra stuff** at the end of the sentence, so there is only one **comma** used to offset it. Can you find them? (There are four.) ___ ___ ___ ___

- Many sentences have dependent clauses that give **necessary information** telling which one, why, when, under what circumstances, or what kind, so they are **not offset with commas**. Can you find them all? (There are eight examples that illustrate this rule.) ___ ___ ___ ___ ___ ___ ___ ___

- Two sentences did not use dependent clauses at all. Can you find them? ___ ___

These two sentences illustrate another punctuation rule: **two independent clauses can be joined with a comma and a conjunction**, not just one or the other, but **both** a comma and conjunction.

This rule looks like this when you have marked the clauses:

___independent___ , and ___independent___	two independent clauses joined with a comma and a conjunction

The next exercise will investigate that further.

CHAPTER 2: EXERCISE 4

Follow these steps for each sentence: 1) Read the sentence, 2) underline the verb(s) twice and the subject(s) once, 3) find the dependent words, which introduce dependent clauses, 4) put the dependent clauses in parentheses, and 5) underline the independent clauses. If two independent clauses are joined by a comma and a conjunction, do not underline the conjunction; it is not a part of either clause and should stand alone.

Ex: Food is one product (that everyone must buy), but many people are worried.

1. Americans are becoming increasingly concerned about the nation's food supply, and many are changing their buying habits.

2. People no longer blindly trust that their food will be safe, for many food scares have occurred.
3. Many instances of *E. coli* contamination have been found in ground beef, and several people have died from eating a bad hamburger.
4. Mad Cow disease outbreaks in foreign-raised beef also make consumers wary of red meat, and many people have become vegetarians in response to their concerns.
5. Even vegetarians are not safe, for spinach was found to be tainted with *E. coli,* as well.
6. After people handle raw chicken, they should wash their hands thoroughly with soap, for *salmonella* can be commonly found on poultry.
7. Children often lick the batter from mixing bowls after their mothers make cakes, yet authorities warn that the raw egg can harbor *salmonella.*
8. The most recent fear involved tainted pet foods, so many dogs and cats died, but no human lives were lost.
9. Consumers are demanding more stringent inspection of foods, or they are choosing not to buy some products.
10. Because many foods are imported from foreign countries, inspection is more difficult, for a food's growing conditions should also be investigated.
11. Organically grown foods, which are grown without pesticides, are gaining popularity, but they are usually much more expensive.
12. Consumers are becoming more knowledgeable about the foods that they buy, yet the fear of buying unsafe products remains.

EXERCISE 4 REVIEW

- Why is there a comma after the word *beef* in sentence #3?

- Which other sentences illustrate this same rule? (You get no hint this time!)

- Why is there a comma after the word *countries* in sentence #10?

- Which other sentences illustrate this same rule?

- What previously studied rule do you see illustrated in sentence #11?

Name ______________________________ Date____________

CHAPTER 2: PROGRESS CHECK QUIZ

Mark the clause structure of each sentence by underlining the independent clauses and putting the dependent clauses within parentheses. If two independent clauses are joined by a comma and a conjunction, do not underline the conjunction; it is not a part of either clause and should stand alone.

Then answer the questions at the end.

1. Because Dan wanted to impress Denise, he rented a limousine to take them to a fancy restaurant.
2. Dan worked part-time at Wendy's and helped his cousin to repair cars when he could find the time.
3. It took six weeks to put aside enough money, but Dan was sure that it would be an evening to remember.
4. After he put on his most impressive suit, Dan called the limousine service, but he was told that he had reserved the limo for the next weekend.
5. Dan was really angry because Denise was waiting for him at that very moment.
6. He did not know where he could find a vehicle to drive.
7. Then Dan remembered the tow truck which was at his cousin's garage.
8. After he jogged three blocks to the garage, Dan was rather sweaty, and his suit was wrinkled.
9. While he muttered a few choice words about the limo service, Dan backed the tow truck out of the driveway, and he drove toward Denise's apartment.
10. Unless her eyes were deceiving her, Denise thought that she saw Dan in a tow truck in her driveway.
11. Dan knocked on the door while he straightened his tie and smoothed back his hair.
12. When Denise opened the door, Dan's eyes widened; Denise looked great!
13. After Dan told her about the mix-up, Denise laughed and said that it did not matter.

14. Dan and Denise went to the fancy restaurant, and the valet parked the tow truck for them.
15. They both had a wonderful time, and they decided to go out the next weekend when the limousine would be available.
16. Because he had spent all of his money, Dan said that they would have to go to McDonald's next time.
17. Denise, who was a good sport, agreed.
18. Dan and Denise had many great times together, but they never forgot the date in the tow truck; it was a memorable evening.

- Why is there a comma after *garage* in sentence #8?

- What other sentences illustrate this same rule?

- Why is there a comma after *time* in sentence #15?

- What other sentences illustrate this same rule?

- Which sentence illustrates the "extra stuff" rule?

CHAPTER 2: EXERCISE 5

Follow these steps for each sentence: 1) Read the sentence, 2) underline the verb(s) twice and the subject(s) once, 3) find the dependent words, which introduce dependent clauses, 4) put the dependent clauses in parentheses, and 5) underline the independent clauses. If there are two independent clauses back-to-back, stop the underlining to show separate clauses.

Ex: Oprah Winfrey came from very humble beginnings; she was raised in Mississippi.

1. Oprah Winfrey has become very popular with American women of all races and classes; she has a hugely popular TV show, magazine, and radio program.
2. Several issues are very important to Oprah; she has repeatedly featured shows on sexual abuse, women's rights, children's issues, literacy, and education.
3. Because she is so respected, many politicians seek an appearance on her show; her opinion evidently has a great influence on her viewers.
4. Many movie stars appear on her show when they have a new movie to promote; the Oprah show provides great advertising and gives viewers a preview of upcoming films.
5. If Oprah likes a movie, it is likely to do fairly well at the box office; featured movies that appeal to women are sure to draw Oprah's viewers.
6. Because Oprah's audience trusts her to be honest, Oprah's greatest asset is her integrity; she vigilantly guards her reputation against any threats.
7. Oprah promotes literacy through her Book Club selections; if she recommends a book, it is sure to be a best seller.
8. One of her Book Club authors made up the details of his book, which was supposed to be based on his true-life struggle with drug addiction.
9. When Oprah discovered that the author was a fraud, she immediately aired the story and apologized to her viewers for endorsing the book.
10. If Oprah ever loses the trust of her viewers, her popularity will fade; however, she is currently considered a trusted friend by millions of females.

EXERCISE 5 REVIEW

- Another punctuation rule states that **two independent clauses having a relationship between their ideas can be joined with a semicolon.** (Note that the second independent clause does not begin with a capital letter; it is still part of the same sentence. A semicolon can take the place of the comma and conjunction to join clauses.)

  This rule looks like this when you have marked the clauses:

  ___independent___ ; ___independent___ two independent clauses joined with a semicolon

- Circle each instance of this rule in the exercise above. Only two sentences do not illustrate this rule. Which are they? ____ ____ (Learn to use semicolons effectively; they make your writing look impressive and show that you understand sentence structure.)
- Why is there a comma after the word *viewers* in sentence #10?

- What other sentences illustrate this same rule?

- Look carefully at sentence #10. This sentence is a little different from the others that also use semicolons. It uses a **transition word to introduce the second independent clause.** What is the transition word? ______________

  This rule looks like this when you have marked the clauses:

  ___independent___ ; transition, independent two independent clauses joined with a semicolon and a transition word followed by a comma

- Using transitions to introduce the second clause adds smoothness and flow to your writing. (Please note that these transition words can be used anywhere in a sentence. *They do not always have a semicolon before them.*) Here is a list of **common transitional expressions that can be used to introduce the second independent clause that follows a semicolon:**

also	**in addition**	**consequently**	**furthermore**
finally	**for example**	**however**	**hence**
indeed	**instead**	**meanwhile**	**moreover**
nevertheless	**then**	**therefore**	**thus**

These expressions usually are followed by a comma when they are used to introduce a clause.

CHAPTER 2: EXERCISE 6

Follow these steps for each sentence: 1) Read the sentence, 2) underline the verb(s) twice and the subject(s) once, 3) find the dependent words, which introduce dependent clauses, 4) put the dependent clauses in parentheses, and 5) underline the independent clauses. If there are two independent clauses back-to-back, stop the underlining to show separate clauses. Include any transition words with the clause that they introduce.

1. Fishing is a great hobby that is enjoyed by people all over the world; in fact, fishing is one of the most popular outdoor activities and is available in most countries.
2. Some people are freshwater fishing enthusiasts; however, others prefer saltwater fishing.
3. Men traditionally have been the sex to enjoy the sport most; consequently, many fishing advertisements feature scantily clad women who are holding fishing rods.
4. However, women are coming on the fishing scene and are winning tournaments; therefore, some companies are changing their ad campaigns.
5. For example, many ads now feature handsome men and decently dressed women who are both enjoying fishing; thus, women are being treated less as sex objects and more as serious anglers.
6. Fishing can be enjoyed from the bank, shore, or dock; therefore, people do not have to invest in a costly water vehicle.
7. A fishing license must be purchased, however, or a person can be fined heavily by the game warden; consequently, most people choose to have one when they fish.
8. Florida has great saltwater fishing opportunities; therefore, many people come from all over to fish in Florida's waters.
9. Some of the close coastline of Florida holds an area that is known as the "flats"; indeed, here the water is from one to six feet deep for miles out and can be navigated in small boats.
10. In this area, one can catch tiny fish or monsters; hence, fishing the flats is exciting.

11. Sometimes dolphins, sea turtles, or stingrays swim in the shallow water; moreover, scallops can be found resting on top of the grass beds which are visible in the clear current.
12. Manufacturers make different baits to fool fish into biting; for example, many lures are made to imitate naturally occurring prey.
13. However, some lures look like nothing that is found in nature; nevertheless, fish still aggressively go for them.
14. Apparently fish have keen eyesight and a good sense of taste; consequently, baits that imitate the look and taste of natural prey are very successful.
15. Artificial baits have some advantages over natural baits; for example, they are often tougher and last longer, and they don't wiggle and squirm as they are hooked.
16. Often artificial baits are very expensive and must be kept cool; nonetheless, they are worth the investment.
17. Fishing is a great family activity; indeed, currently there is a television campaign to increase interest in taking kids fishing.
18. One little girl says to take her fishing because her wedding day will come soon; then a little boy fishes with his father.
19. Kids who are lucky enough to go fishing really love it; consequently, they become adults who have a lifelong hobby that gives them many hours of enjoyment.
20. Fishing is a terrific way to relax, to have fun, and to spend time with family; furthermore, it provides an added benefit: a great dinner!

EXERCISE 6 REVIEW

All of these sentences demonstrate the punctuation rule of **two independent clauses being joined with a semicolon and a transition word that is followed by a comma.** However, four sentences illustrate those transition words also being used in other areas of the sentence. Can you find them? ___ ___ ___ ___

- Why is a comma used after the word *longer* in sentence #15?

- What other sentence illustrates this same rule?

- Why is a comma **not** used after the word *scene* in sentence #4?

- Why is a comma **not** used after the word *adults* in sentence #19?

- Why is *that imitate the look and taste of natural prey* not enclosed within commas?

CHAPTER *three*

Combining Clauses

JUSTIFICATION: WHY DO I NEED TO KNOW THIS?

Being able to combine dependent and independent clauses will enable you to create sophisticated, complex, impressive sentences, a skill which is expected in college-level writing. This ability will also eliminate your making serious sentence errors, errors that deduct major credit from papers.

CLARIFICATIONS

Now that you have successfully completed Chapters One and Two, you should be able to identify the following clause structures:

(dependent) , ___independent___ — introductory dependent clause followed by a comma

___independent___ (dependent) — dependent clause giving necessary information (NO comma)

___independent___ , (dependent) — dependent clause giving extra stuff (= comma)

___independent___ , and ___independent___ — two independent clauses joined with a comma and a conjunction

___independent___ ; ___independent___ — two independent clauses joined with a semicolon

___independent___ ; ___transition___ , ___independent___ — two independent clauses joined with a semicolon and a transition word followed by a comma

Of course, there is always this situation too:

___independent___ . ___independent___ — two independent clauses written as separate sentences

(However, this will result in short, choppy sentences, which are often not advisable in college-level writing.)

Now let's see if you can apply what you've learned!

CHAPTER 3: EXERCISE 1

Does the italicized portion need to use commas or not? Is it necessary or extra stuff? (Which uses commas, necessary or extra stuff?) Add commas where necessary or put **NC** (no commas) next to the sentence number.

1. Backyard bird watching *which requires no expensive equipment* can be a very entertaining pastime.
2. Even birds *that are commonly found in urban areas* can be fascinating to watch.
3. There are a few wild birds *that actually come up very close to me when I am outside.*
4. One mockingbird *which is of a species usually known for being very aggressive* waits for me to put out special foods for him.
5. He flies down immediately to check out any new foods *when I hang up the feeder.*
6. This mockingbird especially likes peanut butter-birdseed balls *that I make especially for him.*
7. When I do yard work, he follows me around to see if I'll stir up any insects *which are the mainstay of the diet of this species.*
8. If I rake leaves, he will fly down to the raked area *after I've finished.*
9. Mockingbirds *which are very territorial* are often seen dive-bombing large birds and animals.
10. I love to hear mockingbirds sing *because their call is so melodious and varied.*
11. Supposedly, a mockingbird imitates the calls of other birds, but I think *that they ad lib their own tunes.*
12. In my yard, another bird *that is also very bold* is a very large ring-necked dove.
13. This dove is so fat *that she looks like a small, grey chicken.*
14. Doves *which are a species of game birds* will commonly freeze their movements if they feel threatened.
15. My friendly dove will fly down very close to me *as I scatter seeds on the ground for her.*

16. She bobs her head as she walks and searches for seeds *which adds to her chicken-like appearance.*
17. Doves are seed eaters and have a lovely call *that sounds soft and mournful.*
18. I often awake mornings to hear the trilling of the mockingbird and the cooing of the dove *sounds that I find wonderful.*
19. Apparently, these two birds do not migrate during winter months *since I see them year-round.*
20. There is also a nest of baby house sparrows in the fern *that is on the front porch.*
21. The babies are covered in fuzzy feathers *which offer little warmth to their tiny naked bodies.*
22. The sparrow parents become very disturbed *when the fern is taken down to be watered.*
23. The male bird has a red-colored head, but the female does not *which makes them easy to distinguish.*
24. Sparrows *which are among the smallest of birds* often nest close to people's houses.
25. Bird-watching is very interesting to me *because I never know what I'll see birds do.*

CHAPTER 3: EXERCISE 2

Create one sentence from the two independent clauses given by making one clause a dependent clause. (Be sure to choose an appropriate dependent word that makes logical sense.) If the dependent clause is at the beginning of the sentence, don't forget to put the comma after it. If it is at the end, is the dependent clause necessary or extra stuff? Does it need a comma?

because
Ex: I am taking tennis this semester ^ I have always wanted to learn to play.

1. Tennis is a popular sport I almost didn't get to take it.
2. The class filled up quickly registration began.
3. I couldn't hit the ball I took the course.
4. The instructor is great he is very popular with students.
5. I found out a certain cute guy was in the class I was glad I enrolled.
6. I had to buy a racket I could play.
7. A cute tennis outfit would be nice the cute guy might notice me.
8. My tennis skirt was lime green it is my favorite color.
9. I first hit the ball it went over the fence.
10. Learning to control my swing was difficult I am not a patient person.
11. The cute guy noticed me we were paired to play doubles.
12. We won our first match he called me for a date.
13. My instructor was impressed I improved so much.
14. The semester was a success I was glad that I took tennis.
15. The cute guy and I are dating tennis brought us together.

CHAPTER 3: EXERCISE 3

Combine the two independent clauses by inserting a comma and an appropriate conjunction (and, but, or, nor, for, yet, so) between them. (Can you tell where they should be divided?)

, and
Example: Food speaks a universal language ^ people all over the world understand it.

1. Cooking shows have high ratings everyone loves to eat.
2. The Food Network is popular millions of people watch it.

3. Some chefs have become celebrities they even have "groupies."
4. Paula Dean is a favorite of men she makes hearty dishes.
5. Rachael Ray has several shows she sells her signature cookware online.
6. Emeril Lagosi does not cook quietly is he shy about his love of food.
7. My mom is a great cook she always looks for new recipes.
8. I can't cook very well I can make simple dishes.
9. My boyfriend loves Mexican food he was raised in Mexico City.
10. He makes great enchiladas and burritos they don't taste like fast food ones.
11. More people are taking cooking classes they are spending a lot of money to learn what some people think is a fading skill.
12. In the past, women learned to cook by helping their mothers in the kitchen most moms nowadays work outside the home and have no time to cook.
13. Many families eat takeout food several times a week some do not eat together at all.
14. Some mothers figure that buying takeout meals is worth the time saved they have no energy to cook when they get home.
15. Many kids prefer the taste of fast food they have become so used to it that homemade dishes taste funny to them.
16. Many men are learning to cook because they have no choice many women find a man who can cook very attractive.
17. Food can be very sensual and sexy it appeals to everyone's basic need to feel satiated.
18. Pleasurable tastes are also satisfying the sounds people make when they taste something really good sound sexual.
19. Favorite dishes bring back happy memories of times spent with loved ones over meals together our memory is related to our sense of taste.
20. Food can be very comforting it has the power to bring people together.

CHAPTER 3: EXERCISE 4

Can a semicolon be used in each of the following sentences? If so, put one and circle it. Should a comma be used instead? If so, put one and circle it, too (make your marks clear and distinct). If the sentence is correct, put **C** next to the sentence number.

1. Coaches can have a great impact on their players they often serve as parental figures to many.
2. Effective coaches push their players to succeed but know when to back off.
3. Good coaches will not reward players for poor effort they will lose the respect of their team.
4. A few words of encouragement that are given when they are truly earned make players feel proud and they work even harder.
5. When players know the coach cares about them personally, they will give all of their effort and do whatever is asked of them.
6. Coaches also serve as role models athletes will follow their example.
7. Some athletes attribute all of their success to the influence of an early coach and the lessons that they learned back then.
8. Athletes respect coaches who are fair, have high expectations, and give emotional support to them for so much of any sport is determined by players' attitudes toward their coaches.
9. Students who choose to become coaches themselves often have a favorite coach who was important to them they even keep in touch over the years.
10. Many kids come from fatherless households or from families who do not spend much time together so coaches become substitute parents for some kids.
11. Like any good parents, coaches must set limits for their athletes and establish clear expectations.
12. Integrity is the most important attribute of truly respected coaches their players must be able to rely on their coach's character.
13. Coaches often serve as counselors and give advice to their players.

14. Many times athletes are motivated by their desire not to disappoint their coach so they will exert all of their effort to playing well.
15. Athletes know the impact their coach has had on them yet the coach may be unaware.
16. Often at end-of-the-year banquets, coaches are rewarded with shiny trophies then they receive an even greater reward.
17. Athletes often thank their coaches for all of their hard work and concern and they may even shed a tear.
18. Coaches have been known to break down too coaching a sport involves a great deal of emotional investment.
19. Playing a sport can be very rewarding coaching one can be too.
20. Coaches often put in longer hours than any classroom teacher they work hard for the sports and the students that they love.

CHAPTER 3: EXERCISE 5

First, read through this exercise as it is written. Does it sound choppy to you? (It should because it uses very few transitions.) Then create one sentence from the two independent clauses by adding a semicolon, an appropriate transition word, a comma, and a lower-case letter.

; however,
Ex. Many students attend college ^ they are not sure why they are going.

1. Most students think that a college degree will guarantee them a high-paying job. That is not always the case.
2. A college education serves to enable students to think for themselves. It changes students' outlooks, viewpoints, thought processes, and maturity levels.
3. Students are required to take core classes in addition to ones in their major area of study. They learn a little bit about a lot of things.
4. College students get to know people of other cultures. Many of their long-held beliefs or prejudices disappear.
5. By taking challenging courses, many students really have to think and study for the first time. They learn to work hard and to apply themselves.

6. Most colleges have an attendance policy. Students must learn to be on time and to appear each day for class.
7. Students become responsible for themselves and their work. They are no longer children.
8. Many college activities require students to cooperate with one another. Students learn social skills.
9. In the classroom, most instructors present both sides of an issue. Students also learn to look at different perspectives.
10. By acquiring a college education, students learn qualities valued by employers. Employees who demonstrate a great work ethic, sound reasoning, reliability, social skills, and intelligence are greatly sought by the working world.
11. College is not trade school. Students do not learn a certain skill that they can sell to the world.
12. They learn to be well-rounded, thinking adults. The value of a college education is in whom students become because of it.

(Now read back over the passage used in this exercise. It should sound smoother with your transitions. However, it is not advisable to use semicolons in every other sentence in your actual writing. Vary your punctuation to show your mastery of clause combinations.)

CHAPTER 3: EXERCISE 6

Correctly combine these independent clauses using the words given. Include any needed punctuation, and be sure that the new sentence makes sense. You may need to change the order of the clauses. (Yes, you must write out the new sentence, but it should be pretty short.)

Ex: I have a cold. I need some Kleenex.

(because) Because I have a cold, I need some Kleenex.
(therefore) I have a cold; therefore, I need some Kleenex.
(and) I have a cold, and I need some Kleenex.

1. My nose is running. I feel awful.

   (and) ______________________________

   (therefore) ______________________________

   (because) ______________________________

2. Chicken soup is good for a cold. My mom is making some.

(since) ____

(so) ____

(consequently) ____

3. I will miss class today. I can't help it.

(but) ____

(however) ____

(although) ____

4. A friend will get my assignments. I will not fall behind.

(therefore) ____

(consequently) ____

(since) ____

5. I e-mailed my instructor. He will know why I'm absent.

(because) ____

(therefore) ____

(as a result) ____

6. I need to sleep. I turned off the television.

(because) ____

(so) ____

(for) ____

7. I will take some medicine. I will rest.

(then) ____

(and) ____

(so that) ____

8. My tissue box is empty. My trashcan is overflowing.

(however) ____

(yet) ____

(although) ____

9. My nose hurts from wiping it so much. I put Vaseline on it.

(moreover) ______________________________

(because) ______________________________

(consequently) ______________________________

10. I feel better. I will go to class.

(when) ______________________________

(therefore) ______________________________

(moreover) ______________________________

Name ____ Date____________

CHAPTER 3: PROGRESS CHECK QUIZ

Correctly combine these independent clauses using the words given. Include any needed punctuation, and be sure that the new sentence makes sense. You may need to change the order of the clauses. (Yes, you must write out the sentence.)

1. My sister can't keep a secret.

   I don't tell her anything important.

   A. because____

   ____

   B. so____

   ____

   C. therefore____

   ____

2. I like spicy foods.

   My favorite restaurant is Jalapeño Joe's.

   A. for____

   ____

   B. since____

   ____

   C. consequently____

   ____

3. Midterms are difficult.

   I know that I'll do well.

   A. yet

   B. although

   C. however

CHAPTER *four*

Sentence Errors

JUSTIFICATION: WHY DO I NEED TO KNOW THIS?

Sentence errors indicate a lack of ability to complete or end sentences correctly. Most instructors consider run-on sentences, comma splices, and sentence fragments to be major errors, and they deduct serious credit from written work that contains them.

CLARIFICATIONS

Remember the Two Truths?

Truth #1: All sentences must have a subject and a verb.
Truth #2: All sentences must make a complete thought.

- "Sentences" without a subject and a complete verb are sentence fragments.

  Examples: The dog with brown spots.
  The dog gnawing on my leg.

No subject and complete verb = fragment

- "Sentences" that are just dependent clauses without an independent clause to complete them are also fragments.

  Example: Because the dog bit me on my leg and caused me to go to the hospital.
  (dependent). = fragment

- A run-on sentence is two independent clauses joined without correct punctuation between them.

  Examples: The dog bit me it hurt badly.
  The dog bit me then it ran.
  The dog bit me but I lived.

___independent___ ___independent___ = run-on

This situation still does not join the independent clauses correctly:

___independent___ (dependent) ___independent___ = run-on

Example: The dog bit me after I petted it then it ran away.

- A comma splice has two independent clauses incorrectly joined with only a comma, instead of using a comma and conjunction or a semicolon.

Examples: The dog bit me, I hollered.
The dog bit me, then it ran.

___independent___ , ___independent___ = comma splice

Refer back to the introduction to Chapter Three to review the correct ways to combine clauses. Use what you've learned about identifying and combining clauses to complete the exercises in this chapter.

CHAPTER 4: EXERCISE 1

Fragments: Identify each as either correct (**C**) or a fragment (**F**). Putting dependent clauses in parentheses and underlining independent clauses can be very helpful if you become confused. Remember, a correct sentence must have at least one independent clause.

_____ 1. Because many people are frightened of them, spiders have a bad reputation.

_____ 2. Even though they are very beneficial.

_____ 3. Spiders eliminate many harmful insects, ones that people should be glad to see dead.

_____ 4. Flies, roaches, moths, and grasshoppers, which are all insect pests.

_____ 5. Spiders are found all over the world and in all kinds of climates.

_____ 6. Some spiders do not spin webs.

_____ 7. Waiting for or jumping on prey.

_____ 8. Many garden spiders catch insects that are attracted to flowers.

_____ 9. After catching the insect and injecting it with venom.

_____ 10. Spiders wrap the prey in their silk to immobilize it.

_____ 11. Hiding it away to eat later.

_____ 12. Spiders do not have to eat very often; their metabolism is slow.

_____ 13. Females that have eaten well will construct an egg sack.

_____ 14. Tiny spiders crawling all over their mother and her web.

_____ 15. Because they look just like little replicas of their mother.

_____ 16. Drafty areas are favorite places to build webs.

_____ 17. Many insects fly along on the breeze and are caught in the sticky web.

_____ 18. Since porch lights attract insects and provide ideal spider habitat.

_____ 19. After an insect's body fluid is consumed.

_____ 20. Spiders tidy up their webs by discarding insect shells.

_____ 21. Until they are disturbed and made to move to another location.

_____ 22. Generations of spiders staying in the same place for years.

_____ 23. Some people suffering from arachnophobia.

_____ 24. Other people don't mind spiders being around.

_____ 25. Because they are helpful in decreasing numbers of pesky insects and won't hurt people unless they are provoked.

CHAPTER 4: EXERCISE 2

Run-ons: Are there two or more improperly joined independent clauses in the following exercises? (Dependent clauses may also be included, but they will not change the fact that the independent clauses must still be joined correctly. Again, marking the different clauses can help clarify the structures that you see.) Put **C** next to the numbers of any correct sentences, and circle the place where the two independent clauses need punctuation in the run-ons.

1. Many people around the world are superstitious and do seemingly strange things to ward off bad luck or to invite good luck.
2. Many people around the world are superstitious and they do seemingly strange things to ward off bad luck or to invite good luck.
3. Some believe that walking under a ladder is to be avoided and some dislike black cats.
4. Many cooks throw a pinch of salt over their shoulders however, some food superstitions are founded on common sense.
5. Baking bread on a rainy day may result in heavy dough the bread will be dense instead of light.
6. My mother-in-law will not wash clothes on New Year's Day because that invites bad luck.

7. On New Year's Day, eating collard greens and black-eyed peas is supposed to bring cash and coins to people I figure it can't hurt.
8. When people move to a new house, they should buy a new mop and broom they don't want to take bad luck along with them.
9. People often make wishes on shooting stars, four-leaf clovers, and dandelion seed heads but do these wishes come true?
10. If you can blow all of the fuzzy seeds off of the dandelion with just one breath, your wish is supposed to come true.
11. Some people will not venture outside on Friday the 13th so they hide indoors.
12. Putting a knife under the mattress of a woman in childbirth will cut her pain and make her labor shorter.
13. Rubbing the head of a red-headed child is supposed to be favorable and warn off bad luck.
14. Rubbing a rabbit's foot is supposed to do the same thing but was it lucky for the rabbit?
15. Are you superstitious or do you think such things are foolish?

Go back and look at the circled areas; what punctuation could fix the run-ons that you've circled? Add the punctuation.

CHAPTER 4: EXERCISE 3

Comma Splices: Are there two or more independent clauses joined with just commas in the following exercises? (Dependent clauses may also be included, but they will not change the fact that the independent clauses must still be joined correctly. Also, remember that commas can be used to add extra stuff to an independent clause. Again, marking the different clauses can help clarify the structures that you see.) Put C next to the numbers of any correct sentences, and for others, circle the place where the two independent clauses are incorrectly joined with a comma splice.

1. Biologists at Atlanta's Georgia Aquarium are taking Dylan home, Dylan is a loggerhead sea turtle that has lived at the aquarium for two years.
2. Loggerheads are an endangered species, and marine biologists are working hard to save them.

3. Dylan came to the Georgia Aquarium after he hatched late, which meant that he stood little chance of making it on his own.
4. Loggerhead sea turtles lay many eggs that normally hatch at the same time, as the baby turtles all race toward the sea together, their chances of survival are greater than if they had hatched individually.
5. Baby turtles are favorite meals of ocean birds and fish, a single baby turtle would quickly be found and gobbled up.
6. Dylan was lucky to have been taken to the Georgia Aquarium, the largest aquarium in the world, where he has received the best of care.
7. Visitors have watched Dylan flapping around in a large tank, a tank that he has shared with many other marine animals.
8. The Georgia Aquarium takes excellent care of its animals and does its best to recreate a natural environment, however, Dylan always looked so sad.
9. Dylan flapped back and forth, appearing to want out of his tank.
10. Colorful fish kept him company, however, there were no other turtles.
11. His eyes always looked forlorn, his expression was always so sad.
12. Happily, however, Dylan is going to be released back on Jekyll Island, the place where he originally hatched.
13. Jekyll Island has a world-renowned turtle hatchery program, which successfully protects and releases many baby loggerheads each year.
14. Some beaches on the island have very strict rules that are designed to protect turtle nests and hatchlings, the turtles return year after year to the same beach where they were hatched.
15. Dylan is definitely large enough now to fend for himself, maybe his own offspring will hatch on Jekyll Island one day.
16. People must work to preserve endangered species, otherwise, Dylan's kind stand no chance of survival.

Go back and look at the circled areas; what punctuation could fix the comma splices that you've circled? Add the punctuation.

CHAPTER 4: EXERCISE 4

Run-ons and Comma Splices: Are there two or more independent clauses joined with nothing or with just commas in the following exercises? (Dependent clauses may also be included, but they will not change the fact that the independent clauses must still be joined correctly. Again, marking the different clauses can help clarify the structures that you see.) Put **C** next to the numbers of any correct sentences, and for others, mark each as either **RO** (Run-on) or **CS** (comma splice) and circle the place where the two independent clauses are incorrectly joined.

_____ 1. Cell phones have changed the way the world works and lives many people find them indispensable.

_____ 2. Cell phones have made instant communication possible, no matter where people are, no longer do people have to wait to talk to someone.

_____ 3. Business transactions can be made via cell phone from anywhere people no longer have to stay at their office.

_____ 4. E-mail and the Internet can be checked via cell phone contact can be maintained no matter where a person is.

_____ 5. Cell phones have changed the social dynamic of people sadly, many families connect to one another through cell phones more so than face-to-face contact.

_____ 6. Many teenagers prefer to text message others instead of talking in person; social skills are weakened as a result.

_____ 7. Because many people are obsessed with appearances, cell phones have become a status symbol the slimmest, sleekest phones are most desirable.

_____ 8. Many additional functions of a phone have had an impact, pictures and videos can be taken at a moment's notice to record events and information.

_____ 9. The chilling cell phone video of the Virginia Tech shootings gave officials valuable information about what actually occurred, many crimes have been solved with the help of cell phones and cell phone records.

_____ 10. The ability to trace cell phone calls and locations can be very useful to law enforcement officials can tell where and when cell phone calls were made.

_____ 11. Many crimes are more easily committed using cell phone technology criminals make extensive use of their phones.

_____ 12. Dishonest students have also discovered many uses of cell phones that enable them to cheat; answers to tests can be sent via text messages notes can be kept on the screen.

_____ 13. In addition, cell phones have become very annoying to others a ringing phone can interrupt a class, a movie, or a conversation.

_____ 14. People who speak loudly on their phones are also annoying some do not care whom they may be disturbing.

_____ 15. Cell phone use can also be a danger to others, talking while driving has been responsible for many wrecks and deaths.

_____ 16. Even using a hands-free device is dangerous the driver still is not fully concentrating on the road.

_____ 17. Drivers who want to make a call should pull over and make the call then.

_____ 18. Talking on a cell phone can put a person in danger, as well; some pedestrians who have been talking on their phones have been hit by cars because they were distracted.

_____ 19. Cell phones have made our lives easier but have they made our lives better?

_____ 20. Cell phone technology is here to stay but people must learn to use it wisely.

Go back and look at the circled areas; what punctuation could fix the areas that you've circled? Add the punctuation.

CHAPTER 4: EXERCISE 5

All Sentence Errors: Identify each as correct (C), a fragment (F), a run-on (RO), or a comma splice (CS).

_____ 1. Every spring, millions of college students head to the nation's beaches to relax, usually they have just finished midterm exams.

_____ 2. Because their exams were so stressful and students feel that they deserve a good time.

_____ 3. Often students cram ten or more people into a motel room that is designed to hold only four, thus they save money on lodging.

_____ 4. Motel owners sometimes claim to have no vacancy they don't want their property destroyed by raucous college kids.

_____ 5. Although students bring revenue to seaside cities and store owners welcome them.

_____ 6. From colleges all across the nation, schools both large and small.

_____ 7. Some students engage in risky behaviors that they ordinarily never would, they may regret this later.

_____ 8. Other students simply go to the beach to relax in the sun and to forget their troubles for a little while.

_____ 9. The sounds of the waves crashing on the beach and of the birds calling in the air.

_____ 10. If they don't want to suffer from sunburn, beachgoers should apply sunscreen several times a day.

_____ 11. Young women worry about looking attractive in a swimsuit, some prepare for spring break for months.

_____ 12. They go to tanning beds and they try to lose weight.

_____ 13. Because shopping for a swimsuit is so difficult and depressing for many young women.

_____ 14. Young men worry, too.

_____ 15. Many young men eat less and begin a weight-lifting schedule to buff up for their beach trip.

_____ 16. Appearance is very important to both men and women.

_____ 17. Since the beach is a place where bodies are on display.

_____ 18. Some students are too shy to appear at the beach in a swimsuit, they choose to go elsewhere.

_____ 19. After their trip to the beach and students return to their classes.

_____ 20. Many students feel refreshed and relaxed after a change of scenery and can apply themselves better to their studies.

Now go back and fix those sentences that had errors. Remember, fragments must have a subject and complete verb and make a complete thought, so you may need to add words or even an entire independent clause to them. Run-ons and comma splices should have correct punctuation added.

Name ____ Date____________

CHAPTER 4: PROGRESS CHECK QUIZ

All Sentence Errors: Identify each as correct (**C**), a fragment (**F**), a run-on (**RO**), or a comma splice (**CS**).

1. _____ Many athletes lift weights as a part of their sports training, coaches require them to log in several hours per week.
2. _____ Because weight training offers several benefits, most athletes don't mind.
3. _____ Although it may take time from other activities and may sometimes be a hassle.
4. _____ Baseball players know that weight training can strengthen arm muscles used for throwing and can strengthen thigh muscles used for squatting.
5. _____ Basketball players can benefit from lifting weights also they can build endurance for long games.
6. _____ Some athletes do not lift weights to build muscle mass they want lean muscles that are strong.
7. _____ Coaches can create a regimen for the desired results, athletes just have to be disciplined enough to do it.
8. _____ Even female athletes benefit from weight lifting, it's not just for men.
9. _____ When women weight train and try to lift something that is too heavy.
10. _____ They must be careful.
11. _____ Anyone can be injured very seriously in the weight room, so athletes should never lift alone.
12. _____ People have died from being trapped under a weight that they couldn't remove themselves which had fallen on their chest or neck.
13. _____ Weight lifters should always have spotters to help them, just in case of an accident.
14. _____ Many feet have been seriously injured by falling weights, so athletes must be wary.

15. _____ Specially designed belts are recommended to help prevent injury people lifting heavy weights should wear them.

16. _____ Pulled muscles can be very painful, thus athletes should follow proper form when lifting.

17. _____ Permanent injury can result from improper lifting; however, proper technique can alleviate this.

18. _____ Weight training can benefit people of any age and light lifting has been recommended for older women.

19. _____ Women prone to the bone disease osteoporosis can slow the progress of the disease they don't have to be hunchbacked.

20. _____ Weight training is beneficial if done correctly, therefore, more people should consider adopting a weight-lifting program.

CHAPTER *five*

Commas

JUSTIFICATION: WHY DO I NEED TO KNOW THIS?

Commas are considered by many people to be the most commonly misused type of punctuation, and to illustrate mastery of the use of commas is the mark of a competent writer. You do want to be viewed by others as competent, right?

CLARIFICATIONS

The good news is that you have already learned the most difficult comma rules from the preceding chapters and that there are really only seven comma rules!

Rule 1: Between independent clauses with conjunction

(Hint: Look for subject and verb after conjunction.)
Ex: We asked for a raise in our allowances, and my dad agreed.
He should have known better, but he was taken in by our sweet smiles.
We should be ashamed and should offer an apology. (Why is there no comma here?)

Rule 2: Essential/non-essential (necessary/extra stuff)

("Extra stuff" needs commas; necessary information needs no comma.)
Ex: The chicken pox, which is a highly contagious disease, is spreading all over Georgia.
(Which disease? The chicken pox. Extra stuff? Oh, it is highly contagious, too.)
Ex: A disease that is highly contagious can spread rapidly.
(Which disease? A disease that is highly contagious… Ah, necessary stuff!)

Rule 3: Introductory elements (The trickiest one you know already!)

Introductory dependent clause
Ex: After Millie closed the door, she covered her ears and screamed loudly.

3+ words in a prepositional phrase = ,
Ex: After the long train ride, Emily felt exhausted.

Conversational elements
Ex: Yes, I'll be happy to do that for you.
No, I wouldn't ever do that!
Well, I'll have to think about it.

Participial phrase (looks kind of like a verb, but acts like an adjective)
Ex: Studying every day, the student brought his average up.
Worried and afraid, he slowly opened the envelope containing his grades.

Rule 4: Interrupters (These need two commas.)

Appositives (These rename the preceding noun, which sounds like extra stuff to me!)
Ex: Sam Delany, a freshman college student, made the Dean's List.

Transitional expressions (Remember these? They can occur anywhere in a sentence!)
Ex: Sam did not, however, have a cumulative 4.0 GPA.
I, on the other hand, do not eat seafood.

Contrasting expressions
Ex: Sam, not Sally, will go with the team.

Rule 5: Items in a series

Ex: Chicken, fries, and slaw were served. (Do not separate the subject from the verb with a comma unless you have a very good reason.)
Dan liked to swim, jog, and box to keep fit.
I love my mom, I respect my dad, but I despise my brother.

Rule 6: Two + adjectives before a noun

(Hint: Try switching the adjectives around or using "and" in between them to see if the adjectives modify equally. If they do, put a comma; if they don't, they aren't equal, so you don't put a comma.)
Ex: It was a beautiful, sunny fall day. (Use no comma between sunny and fall because "sunny and fall day" sounds weird.)
Her long, lustrous auburn hair was her best feature. ("Lustrous and auburn hair" sounds weird so no comma.)

Rule 7: Conventional situations

Dates
Ex: I was born on Tuesday, April 28, 1992, and lived to be a ripe old age.
(Note the comma after 1992. Three or more parts to a date will have the last part followed with a comma, no matter what that part is. Sorry, I just explain these rules; I didn't create them.)
Ex: I was born on April 28 and was baptized on June 3.
(Why no comma here?)

Addresses
Ex: I was to send it to Sam Taylor, P.O. Box 234, Dawson, Georgia 31742. (Use commas between all the parts of an address—think of addressing an envelope—but use no comma after the state if it is followed by a zip code.)

Ex: Sam has lived in Dawson, Georgia, all of his life. (Use a comma between the city and the state—or city and country—and after the state if there is no zip code afterward.)
Cheryl was born in London, England, and later moved to America.

Titles/degrees
Ex: Finara Newman, D.D.S., will be our speaker at the meeting. (The information about her degree is like extra stuff.)

Nouns of address (Speaking directly to someone and addressing that person by name or title requires the name or title to be offset by commas.)
Ex: I told you, Dave, not to do that anymore.
Sherry, will you hand me that pencil?
You are fired, sir.
I quit, you jackass!

Quoted conversation (This is covered in greater detail in Chapter Seven: Quotations and Italics.) Commas are used to offset actual spoken quotes from the dialogue tags.

Ex:

Jack announced to the group, "I am moving to New York City next month, so I'll need someone to take over for me at the firm."

"What? How could you? I won't let you go without me," Rhonda dramatically responded. "Take me with you!"

"No, I'm going alone, Rhonda," Jack replied, "because there won't be enough room in the car for you and my stuff too."

Rule 8: Don't use unnecessary commas.

(If you can't think of a rule that applies, don't put a comma!)

Illustrate your ability to analyze the structures in each sentence and apply these simple rules in the following exercises.

CHAPTER 5: EXERCISE 1

Insert commas where they are needed and circle them. Ask yourself which comma rule applies to each comma to test whether you really need it. Be prepared to defend your decision.

1. People who are in the market for a used car must investigate their choices carefully or they will be sorry later.

2. Since there are so many car dealers and private owners wanting to make a sale shoppers must be wary.
3. In order to make a wise choice shoppers should have a clear idea ahead of time of what they want what they can afford and what to look for in a vehicle.
4. Because looks can be deceiving a perspective buyer should check the car's engine and history or hire a mechanic to check out the car.
5. If the previous owner kept records of each time the car was serviced the buyer can be reassured that the car was maintained properly and that the engine should be in good shape.
6. Although one should expect a used car to have been driven buying a car with very high mileage may be a bad idea.
7. However some cars with very high mileages may still be good deals depending on the price and the expected life of the engine.
8. Many imports have higher expected engine lives than domestic American cars so shoppers should consider this in their deliberations.
9. Because people who keep their car's interior clean are more apt to keep their car well maintained overall the condition of the interior can be an indicator of a car's past upkeep.
10. A car whose previous owner smoked will be difficult to clean well enough to remove the odor so non-smoking shoppers may want to ask about this before even viewing the car.
11. Shoppers have many options for comparing cars such as shopping car dealerships online viewing newspaper advertisements and visiting car lots.
12. Because some salesmen can be annoying many people shop car lots after the business has closed for the day.
13. Many used cars have been through floods and may have sustained hidden damage but shoppers can look for telltale signs if they know how.
14. If mud is found under the trunk carpeting or lining the buyer should take this as an indicator of flood damage.

15. By purchasing a car that is only a year or two old a buyer can take advantage of big savings in the depreciated value of the car yet he or she still gets a car that is relatively new.
16. Some people refuse to buy new cars because they depreciate immediately after the purchase so some always shop for slightly used vehicles.
17. Considering the expense of a vehicle even used-vehicle shoppers must be careful.
18. However by investigating used cars carefully buyers can find great vehicles at fantastic prices.

EXERCISE 1 REVIEW

- What two comma rules are found in sentence #3?
- What two comma rules are found in sentence #7?
- What two comma rules are found in sentence #11?
- What two comma rules are found in sentence #15?
- What one comma rule is found in sentence #17?

CHAPTER 5: EXERCISE 2

Insert commas where they are needed and circle them. Ask yourself which comma rule applies to each comma to test whether you really need it. Be prepared to defend your decision.

1. Although many people are skeptics ghost stories are a part of every culture's folklore.
2. In literature from all over the globe one can find ghosts goblins spirits and the "undead."
3. In Korean folklore a bathroom ghost asks visitors if they prefer red or blue toilet paper and their answer determines whether they are killed or let go.
4. Koreans are also familiar with the Sapsal dog a dog taken into battle to chase vengeful spirits away from living soldiers.
5. The members of one African tribe the Betsimisaraka are so afraid of disturbing the ghosts of the dead that they will not talk loudly laugh or walk near a grave for fear of angering ghosts.
6. Some African ghosts are known as kinoly which look like living people who have red eyes long fingernails and a nasty habit of disemboweling people.
7. In the folklore of the Native American nation of the Ojibwa a terrifying creature lurks one with a lipless mouth jagged teeth and a hissing voice.
8. The Ojibwa call it the Windigo and they tell stories of its eating anyone who ventures near it and leaving blood-filled footprints behind.
9. Ghost stories abound about the Tower of London the infamous English place of execution for many prisoners and today people can take a ghost tour of the 900-year-old building.
10. Sir Walter Raleigh chained and headless has been seen walking the ramparts near where he was kept prisoner.
11. Another headless apparition of the Tower is Anne Boleyn second wife of Henry VIII mother of Elizabeth I and unfortunate past prisoner of the Tower.
12. In a Japanese ghost story called "The Story of Okiku" Okiku a maid in the home of a samurai named Aoyama accidentally breaks several precious ceramic plates treasures of the family.

13. When Aoyama discovers this he kills Okiku in a fit of rage and throws her corpse in a well.
14. However every night thereafter Okiku's ghost rises from the well counts the remaining plates and breaks into loud sobs eventually driving Aoyama insane.
15. A Russian story of the supernatural relates the tale of a soldier who is chased by two corpses but he eventually escapes when they fight over which one will eat him.
16. There are many ghost stories set in the American South and one that is quite famous in Cuthbert Georgia is that of Peg Leg Pete a soldier who was wounded in the Civil War.
17. Peg Leg Pete is said to haunt Old Main the present administration building of historic Andrew College founded in 1854.
18. Over the years students have reported seeing and hearing Peg Leg Pete as he hobbles through the hallways of Old Main which was built on the site of a Civil War hospital.
19. Since ghost stories apparently are universally found in all cultures do you think that there may be something to them?
20. Have you ever had an encounter with a ghost spirit or apparition?
21. Most people who are skeptics have never experienced the supernatural but once they do they may have to leave their skepticism behind.

EXERCISE 2 REVIEW

- What two comma rules are found in sentence #2?

- What two comma rules are found in sentence #5?

- What two comma rules are found in sentence #9?

- What two comma rules are found in sentence #10?

- What three comma rules are found in sentence #16?

CHAPTER 5: EXERCISE 3

Insert commas where they are needed and circle them. Ask yourself which comma rule applies to each comma to test whether you really need it. Be prepared to defend your decision.

1. Because I'd never been to a real concert I told my mom that I wanted tickets to a My Chemical Romance concert as my birthday present.
2. Since she knew how much I loved the group she agreed and we searched the Internet for concert dates locations and ticket prices.
3. Of course I didn't want to go alone and my mom didn't want me to either so we looked for prices for two tickets.
4. While searching for tickets on MySpace.com we saw tickets for lots of great bands such as Emory Underoath and Linkin Park.
5. We finally found the best deal on tickets however through TicketMaster.com and bought two for the performance at Hi Fi Buy's Amphitheater.
6. I took note of the Linkin Park concert that would be at the Tabernacle in January and I decided to ask for those tickets for Christmas.
7. When I told my buddy Tom that I had tickets to see My Chemical Romance at the Amphitheater he thought that that was cool.
8. However when I asked him if he'd like to go too he freaked!
9. For weeks before the concert we made plans for that night and talked about how much fun we would have.

10. Finally the night of the concert arrived but as I was about to leave the house to pick up Tom the phone rang.
11. Tom was moaning and groaning about how he had a fever diarrhea and apparently a stomach virus so he couldn't go but hoped that I could find someone else to take his place.
12. Because he was so upset and pitiful I didn't want to make him feel worse so I told him that I'd tell him all about the concert the next day.
13. As I hung up the phone I was so mad because I didn't know of anybody who would be able to go with me especially at such short notice.
14. "You'd better get going or you'll be late" my mom said.
15. As I turned and looked at my mom an idea came to me an idea that I never thought that I'd ordinarily consider.
16. As we rode to the Amphitheater my mom looked pretty good in her jean jacket Gap t-shirt and blue jeans and she was excited to be asked to go.
17. She told me about the concerts that she had been to before Tina Turner the Rolling Stones the Beach Boys and Chicago groups that I'd heard of and actually liked a little.
18. Holding hands we pushed through the crowd to find spaces close to the stage and we joined the mosh pit of bodies that were bopping up and down to the beat.
19. Nobody in the crowd seemed to think twice about my standing with a woman who was obviously old enough to be my mother and they smiled at her as she bopped up and down with everybody else.
20. We became quite popular with the people standing close by us as we laughed and danced together and when they found out that she was my mom they liked us even more.
21. The highlight of my evening however was when I was lifted to the top of the crowd to surf along the hands of the audience and as I looked to the side I saw my mom surfing along too laughing and smiling like a teenager.
22. Wow I'd have this birthday memory forever one that included my great mom.

EXERCISE 3 REVIEW

- What three comma rules are found in sentence #2?

- What three comma rules are found in sentence #4?

- What two comma rules are found in sentence #12?

- What three comma rules are found in sentence #16?

- What four comma rules are found in sentence #21?

CHAPTER 5: EXERCISE 4

Insert commas where they are needed and circle them. Ask yourself which comma rule applies to each comma to test whether you really need it. Be prepared to defend your decision.

1. Bill-paying time is here again and I must take care of this horrid depressing time-consuming chore.
2. Before Tuesday June 5 2007 the credit card payment must be sent to USAA Savings Bank P.O. Box 14365 Las Vegas NV 89114.
3. I must also send the electric bill payment to Georgia Power 241 Ralph McGill Blvd. Atlanta GA 30308 before June 16 2007 or be charged a late fee.
4. Because funds are limited I must stagger monthly payments as I get paid or I'll be in big trouble with my bank Bank of America.

5. Keeping up with what needs to be paid by what date is a chore but I have to do it.
6. Jane would you take over this wretched awful aggravating task for me?
7. Since Jane refuses Todd will you do it?
8. Because both Jane and Todd are close friends of mine I'm seriously considering this especially since they are both accountants.
9. Their accounting office is located at 241 Baker Street Cuthbert GA 39840 and the business is doing quite well.
10. Last year they opened their new business Money Matters on Wednesday January 4 2006 to take advantage of the tax breaks offered by the city government.
11. If I could pile my bills in a box give Jane and Todd my account numbers and have them pay my bills for me I'd be so happy!
12. That's enough daydreaming Self so get back to work!
13. The bill to Animas Corporation must be mailed to P.O. Box 2048 West Chester PA 19834 by Thursday May 31 and will pay off that account.
14. Also Dr. Neil Schaffner M.D. must be paid $100 which applies to the amount of my insurance deductible.
15. Insurance is so expensive but I'm thankful that I have coverage coverage that protects me from catastrophic costs.
16. The State Farm payment for the car insurance must be paid by Friday June 15 and will not be due again until September.
17. My Verizon bill is due by June 16 2007 but includes two months' worth of service.
18. Todd please think about taking over my bill paying or Jane if he won't do it I'll trade you my housecleaning service for your accounting skills.
19. What do you say Todd and Jane?
20. Well I guess that I will have to get used to paying my own bills so I'd better get to it.

EXERCISE 4 REVIEW

- What two comma rules are found in sentence #1?

- What three comma rules are found in sentence #4?

- What two comma rules are found in sentence #6?

- What three comma rules are found in sentence #14?

- What four comma rules are found in sentence #18?

CHAPTER 5: EXERCISE 5

Insert commas where they are needed and circle them. Ask yourself which comma rule applies to each comma to test whether you really need it. Be prepared to defend your decision.

1. Online classes can be really convenient for students but they aren't for everybody and students should consider their own strengths and weaknesses before registering for one.
2. For adult students who have jobs or children to take care of online classes can be great for their schedules.
3. Instead of having to find course offerings that don't interfere with work or to arrange for childcare non-traditional students those older than most college students can take online classes which allow them to study at night or stay home with children.
4. In addition online classes allow students to take classes from home which saves students travel time and gas money a real cost savings nowadays.
5. Many students however mistakenly think that online courses are easier than traditional courses but nothing could be further from the truth.

6. Because students choose when to do most coursework for online classes successful students must be self-disciplined motivated and mature.
7. Otherwise students procrastinate thinking that they have plenty of time to do assignments and usually failing to do a good job if they do it at all.
8. Online students must also be computer savvy able to send receive and create documents effectively for submission to the instructor.
9. In taking online courses students basically are in charge of educating themselves learning the material on their own which requires excellent reading comprehension skills.
10. Of course online students may submit questions to their instructor but some instructors do not bother answering them.
11. When taking online classes students are often required to take online tests a process which can be very stressful for students.
12. If technology glitches occur online students can become very frustrated and worried as to whether their instructor knows about the problems.
13. Surprisingly most college instructors are not trained in the creation of online courses and therefore their classes my not be very well structured.
14. Because of the popularity of online offerings most textbook companies offer online versions of their textbooks' exercises and compatible course materials.
15. Instead of creating their own online courses many instructors choose to adopt textbooks that offer courses that are prepackaged and come with the student text a choice that proves expensive for the student.
16. Ideally any instructor who teaches online should have had plenty of experience in taking online courses thereby learning what works and what doesn't from the student's perspective.
17. Many students think that instructors who teach online have an easier time than those who teach face-to-face classes yet again they are mistaken for teaching online classes takes much more time in preparing the material for the course and responding to the students' questions and assignments.

18. If you ever decide to take an online course make sure that you are mature enough to make yourself do the work are able to learn on your own and can operate a computer fairly well.
19. Furthermore going to meet the instructor in person at the start of the course is a good idea so that you have an idea of the person behind the computer.
20. Finally prioritize your schedule making completion of your online coursework a major objective every day even on the weekends and by doing this you will be a successful online student.

EXERCISE 5 REVIEW

- What one comma rule is found in sentence #1?
- What two comma rules are found in sentence #5?
- What two comma rules are found in sentence #8?
- What two comma rules are found in sentence #17?
- What four comma rules are found in sentence #20?

Name ______________________________ Date__________

CHAPTER 5: PROGRESS CHECK QUIZ

Insert commas where they are needed and circle them. Ask yourself which comma rule applies to each comma to test whether you really need it. Be prepared to defend your decision.

1. Please driver take me to the office of Dr. Max Hoffer D.V.M. which is located at 135 South Suffolk Street South Bronx NY 89347.
2. Because it was snowing lightly outside Rose unused to such weather put on a heavy coat wool scarf fur-lined hat and snow boots.
3. The date of Wednesday April 18 2007 will live in infamy because that's the day I got my driver's license my brother's old car and my first speeding ticket.
4. I wish Robert that you would listen to what I say even if you don't want to!
5. On the other hand if children are allowed to have anything that they want not only will parents be broke but also children will never be satisfied.
6. Yes Phyllis you karaoke queen I'll go with you but I won't sing no matter how many beers I have!
7. Walking four miles from the car John finally found a gas station one that fortunately was open at 3:00 a.m.

QUIZ REVIEW

- What three comma rules are found in sentence #2?
- What three comma rules are found in sentence #5?

CHAPTER *six*

Semicolons and Colons, Parentheses and Dashes

JUSTIFICATION: WHY DO I NEED TO KNOW THIS?

Being able to use semicolons, colons, parentheses, and dashes effectively allows you to vary your punctuation, offers more flexibility in structuring your ideas, and empowers you to precisely express yourself.

CLARIFICATIONS

Semicolons

In Chapter 2, you learned the most common use of the semicolon, to join two independent clauses:

<u>independent</u> ; <u>independent</u> two independent clauses joined with a semicolon

You also learned a slight variation of that rule, to join two independent clauses with the second clause having an introductory transition word followed by a comma:

<u>independent</u> ; <u>transition , independent</u> two independent clauses joined with a semicolon and a transition word followed by a comma

One more use of the semicolon is to use them in place of commas when individual items in a series already contain commas. After all, the reader can't clearly see each item because of all of the confusing commas.

<u>Confusing:</u> (What goes with whom?)
I met lots of interesting people at the cocktail party: Dr. Lela Phillips, author of *The Lena Baker Story*, Karan Pittman, genealogist extraordinaire, Dr. Sherri Taylor, bounty hunter impersonator, Dr. Susan Tusing, Fulbright Scholar, Linda Grice, Auburn fanatic, and Yamandu Acosta, world traveler and epicurean.

Clearer:
I met lots of interesting people at the cocktail party: Dr. Lela Phillips, author of *The Lena Baker Story*; Karan Pittman, genealogist extraordinaire; Dr. Sherri Taylor, bounty hunter impersonator; Dr. Susan Tusing, Fulbright Scholar; Linda Grice, Auburn fanatic; and Yamandu Acosta, world traveler and epicurean.

That's logical, right?

Colons

Colons call attention to what follows them; they seem to say, "Here's what I mean by that" or "Here's an example of that."

- They are most commonly used between two independent clauses, like semicolons.

Example:
Monday night was the most embarrassing night of Shana's life: she unknowingly walked out of the ladies room with her skirt tucked into her pantyhose, and everyone noticed before she did.

- Colons can also draw attention to single words or phrases that serve as examples of what was mentioned.

Example:
Studies have proven that there is one factor that is the most influential in keeping people out of poverty: not having children outside of a committed marriage.

- Colons are often used in research papers to introduce quoted passages or paraphrases from outside sources that prove the point the author is making.

Author's point > **Flocking birds often signal schools of fish to fishermen:**
Quoted Passage> **"Captain Pete Rose lives by these birds when he's searching for yellowfin tuna off the Bahamas" (Poveromo 35).**

From George Poveromo's article entitled "Air Support" in *Saltwater Sportsman*, May 2007.

- Colons can introduce items in a series. However, there is one quirky little rule to remember concerning colons: they cannot be used after a verb or a preposition.

Wrong:
(verb)
The list of celebrities **included:** Jamie Foxx, Robin Williams, Queen Latifa, Tom Hanks, and Denzel Washington.

Correct:
The list of celebrities included Jamie Foxx, Robin Williams, Queen Latifa, Tom Hanks, and Denzel Washington.

Wrong:

(preposition)

Many celebrities were present, such **as:** Jamie Foxx, Robin Williams, Queen Latifa, Tom Hanks, and Denzel Washington.

Correct:
Many celebrities were present, such as Jamie Foxx, Robin Williams, Queen Latifa, Tom Hanks, and Denzel Washington.

- Colons also are used to make impressive titles for essays, papers, books, etc. Start with a generalization, put a colon, and then give a specific about it:

Paris Hilton: Debutante in Danger of Destruction
School Reform: Smaller Is Better
MTV: Its Influence on My Generation

(Try this with your next paper! It looks very impressive!)

CHAPTER 6: EXERCISE 1

Add semicolons or colons where necessary. If no mark is needed, put a **C** next to the number of the sentence. Be ready to defend your choice.

1. My Walmart stock rose two points my Whirlpool stock fell three points.
2. Survival reality shows have become quite popular they pique the curiosity of everyone who wonders about their own survival skills.
3. Every woman should have three things in her purse at all times Kleenex, pain reliever, and a good neutral-color lipstick.
4. Meg learned to cook from her mother she learned to diet from Weight Watchers.
5. An apple a day keeps the doctors away a good insurance plan brings them out of the woodwork.
6. Vince got the best piece of advice from his dad don't lend money to a friend unless you don't care whether you get it back.
7. Sydney's beach bag was filled with a towel, sun block, three diet Cokes, and a bag of Doritos.
8. Nan couldn't believe what had happened to her her car had been buried under an avalanche of snow.

9. We will need the following items bug spray, sleeping bags, matches, drinks, and sandwiches.
10. Ted wanted to play basketball so badly his GPA made him ineligible, however.

EXERCISE 1 REVIEW

- Which sentence was correct already?

- What mark did you use in sentence #3? Why?

- What mark did you use in sentence #4? Why?

CHAPTER 6: EXERCISE 2

Add semicolons or colons where necessary. If no mark is needed, put a **C** next to the number of the sentence. Be ready to defend your choice.

1. My favorite movie is Mel
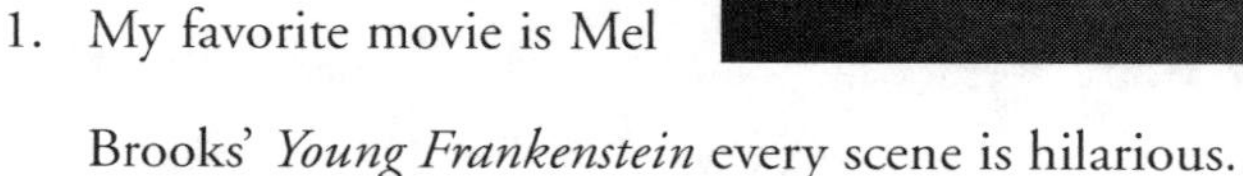
Brooks' *Young Frankenstein* every scene is hilarious.
2. The movie is a comedy based on the book entitled *Frankenstein*, by Mary Shelley filmed in black and white, it parodies the early horror movie genre.
3. The original 1931 horror movie starred Boris Karloff as the monster the 1974 Mel Brooks comedy features Peter Boyle as the monster.
4. *Young Frankenstein* incorporates a cast of comedy greats Gene Wilder, Marty Feldman, Cloris Leachman, and Madeline Kahn.
5. Marty Feldman, a bug-eyed British actor, plays Igor, the hunchbacked assistant Feldman drove the camera crew crazy by switching his hump from one side of his back to the other.

6. Cloris Leachman has a line that never fails to crack me up in a very thick German accent, she says very seriously, "Stay close to de candles; de staircase can be treacherous," but the candles aren't even lit!
7. Igor procures the body parts for building the monster however, he mistakenly picks out a brain labeled "Abnormal."
8. The movie includes many sexual innuendoes references to "a roll in the hay," big knockers, and speculation on the size of the monster's genitals keep the audience laughing.
9. The inarticulate monster grunts and groans as he tries to communicate however, after a sexual encounter with him, Madeline Kahn hilariously bursts into operatic song during her throes of ecstasy.
10. Gene Hackman makes a brief appearance as the blind man he offers a cigar to the monster, but instead of lighting the tip, he lights the monster's thumb.
11. Brooks' movie uses props from the 1931 movie, such as the electrical gizmos, laboratory equipment, operating table, and bubbling tubing.
12. One highlight features a tap dancing routine with Dr. Frankenstein and the monster they dance to the song "Puttin' on the Ritz."
13. The movie ends with a daring experiment the monster and his creator undergo a brain wave transfer.
14. They each get something from the other the monster can speak intelligently, and Dr. Frankenstein apparently grows larger genitals.
15. I love this movie it never grows old, and it's always funny.

EXERCISE 2 REVIEW

- Which sentence was correct already?

- What mark did you use in sentence #3? Why?

- What mark did you use in sentence #4? Why?

- What mark did you use in sentence #6? Why?

- What mark did you use in sentence #9? Why?

- What mark did you use in sentence #13? Why?

CHAPTER 6: EXERCISE 3

Add semicolons or colons where necessary. Some may need both. If no mark is needed, put a **C** next to the number of the sentence. Be ready to defend your choice.

1. The following textbooks were approved by the Board *Our Changing World* grade 6 *Social Studies and You* grade 7 and *Cultures across the Globe* grade 8.
2. The Superintendent announced raises for eligible employees "About fifty employees are expected to be the beneficiaries of a $1,000 pay hike," said Superintendent Smith.
3. We need to have all of these business owners present for the Chamber of Commerce meeting Ralph Lamb Just for Ewe Chris Stone Pavement Pros Paul Pace Carland and Shannon Scrubb Magnificent Maids.
4. My agriculture paper was entitled "Drought The Economic Devastation from Rain Scarcity."
5. Many cities experienced high temperatures Miami 97 Atlanta 94 Augusta 92 Birmingham 94 and Memphis 93.
6. Students earning Honor Roll awards are Tom Mote, Didi Sessions, Mandy House, Katie Campbell, and Dan Bane.
7. The president of the local chapter of the NAACP announced a number of awards would be made at the meeting "At the reception, we will honor those who have made outstanding contributions during the past year, and we will award several scholarships to local students."

8. Missy's thesis for her class was entitled "Mark Twain Portrait of a Misogynist."
9. Richard found some great deals at the flea market a Penn reel with fiberglass rod $30 a pair of genuine snakeskin boots $20 a vintage guitar $50 and a 1980 Nikon camera $45.
10. John finished his criminal justice paper, which he called "The C.S.I. Craze A Reflection of American Society."

EXERCISE 3 REVIEW

- Which sentence was correct already?
- What marks did you use in sentence #3? Why?
- What mark did you use in sentence #7? Why?

MORE CLARIFICATIONS

Parentheses

- Parentheses are used to set off and **de-emphasize** (rather offhandedly mention) extra stuff.

Example: My mother (regretfully I must call her that) deserted me when I was born.

Note: Commas could be used here too, but they call more emphasis to the extra stuff than parentheses:

Example: My mother, regretfully I must call her that, deserted me when I was born.

It all depends on how the author wants the material to be presented. How much importance should be put on the material?

- Parentheses are also used to enclose explanatory matter:

Author's point >	**Flocking birds often signal schools of fish to fishermen:**
Quoted Passage>	**"Captain Pete Rose lives by these birds when he's searching for yellowfin tuna off the Bahamas" (Poveromo 35).**

The parentheses explain that the quote was from an author named Poveromo and found on page 35. (The reader would check the Works Cited page to see the title of the work.)

Dashes

- Dashes are occasionally used to create an abrupt interruption in thought or to call **emphasis** to something. Dashes are typed by typing two hyphens.

Example: My mother — I have no other term, unfortunately, to describe the woman who deserted me when I was born—had never been a part of my life.

CHAPTER 6: EXERCISE 4

Add parentheses or dashes where necessary. (In several of these, a colon or commas could be used, but just stick to dashes and parentheses.) If no mark is needed, put a **C** next to the number of the sentence. Be ready to defend your choice.

1. The dog a beautiful Golden Retriever that weighed about 120 lbs. was originally a gift from his grandfather.
2. Don't open the pot's lid unless you really want third degree burns from the steam.
3. Jen estimated that she had been asleep as close as she could tell from the position of the sun for about three hours.
4. The piano a magnificent Yamaha grand piano made Alice's heart skip a beat.
5. Even though the steaks were cooked to perfection, Anne would not eat any she had been a vegetarian for years.
6. Sundays were for family gatherings everyone would come to Grandma's house after church and bring a covered dish.
7. The t-shirts black shirts with gold lettering would take at least three days to create.
8. Lon was sure that he had seen something whether it was the Loch Ness Monster, he couldn't say.

9. The course syllabus a lengthy document of five pages clearly specified everything that was expected of the students.
10. Tonya's art class a three-hour-long art class was exhausting and left her very tired.
11. To be named valedictorian of her class was a goal Latisha never lost sight of she made her schoolwork her first priority.
12. The wellness class a requirement for all students was filled to capacity.
13. No matter what happens, do not do not press this button!
14. Sandra liked her aerobics class especially since the instructor was really cute.
15. Yoshiro wanted to own his own business and set his own hours.
16. The most important thing to remember and I'll remind you of this everyday is to securely lock the door after you arrive.
17. Since the foundation is crumbling, someone will have to go under the house to set the supports but I'm not volunteering!
18. Many people believe that deja vu feeling that one has experienced something before is proof of reincarnation.
19. The air conditioner one that was at least thirty years old was so energy inefficient that the family could not pay their first electric bill.
20. If students do not complete the paperwork all of the paperwork for the loan, it can not be processed.

EXERCISE 4 REVIEW

- Which sentences could be correct already?

- What marks did you use in sentence #7? Why?

- What mark did you use in sentence #13? Why?

Name ______________________________ Date__________

CHAPTER 6: PROGRESS CHECK QUIZ

Add semicolons or colons where necessary. Make your marks clear and distinct.

1. Ronnie and Sherryl went to see the Dean of Student Life they were very unhappy with the cafeteria food, which they were sure was making them sick.
2. What do you want to do tomorrow play basketball, go to the movies, or shop for groceries?
3. I could choose from several different CDs Taking the Long Way, by the Dixie Chicks Get Rich or Die Trying, by 50 Cent Come Away with Me, by Norah Jones or The Open Door, by Evanesence.
4. According to Stan Smith, a reporter for WALB News, the crime was a violent one "We could see blood-spattered walls when the emergency crew opened the door of the residence."
5. Many people are afraid of stormy weather surviving destructive storms will do that to people.

Add parentheses or dashes where necessary. Make your marks clear and distinct.

1. Trina's divorce an ugly, very public one was fodder for gossip columnists for months.
2. Because they wanted more of a voice in the discipline policy, many students attended the meeting the room was packed.
3. I will never never go out with him again!
4. Most of the dogs strays that had been abandoned at the shelter had fleas when they arrived.
5. Tasha's new computer a Gateway laptop made completing her schoolwork so much easier.

CHAPTER *seven*

Quotations and Italics

JUSTIFICATION: WHY DO I NEED TO KNOW THIS?

Quotations marks and italics are used quite often in college writing, for research papers and narratives especially, and a competent writer must know how to use them.

CLARIFICATIONS

Quotations

- Quotation marks are often used in narratives to show conversation:

"I do not understand what this woman wants us to do!" Teresa wailed as she threw down her pen in frustration.

"What in the world is the problem?" asked Shakela, her roommate. "What's the assignment?"

"We are supposed to write a hypothesis—no, wait, several of them—of why an experiment turned out the way it did," Teresa whined. "I don't know how she wants us to write this friggin' report! I can't do this!"

"Hold on! Hold on!" Shakela said calmly. "I had her for that class last year, and I can help you. Let me see the assignment sheet."

While Shakela read the assignment, Teresa sat with her arms folded and pouted.

"Okay, I see. It's pretty easy. You can do this," Shakela said.

"No, I can't," Teresa pouted, "and that woman's crazy if she thinks I'm goin' to do this report."

"Well, with an attitude like that, you sure can't do this assignment," Shakela scolded calmly. "You can't do anything with such a negative attitude. Do you want my help or not?"

Teresa sulked quietly for a minute.

"Yeah, I'd like your help. I'm sorry. I guess I need to develop some patience," Teresa said, with a sheepish smile.

"Good. Let's get to work," Shakela responded.

"Okay," said Teresa.

Notice a few rules for writing dialogue:

1. Each person's conversation is indented as a new paragraph, no matter how little or much is said.
2. Quotation marks enclose only the person's exact conversation; narrative, storytelling elements are excluded.
3. Dialogue tags, the "said Teresa" or "Shakela responded," are included to tell the reader who is speaking, and they often tell how the dialogue was said or what was going on at the time.
4. Quoted sentences are punctuated as usual. However, people often speak in fragments, and so long as the fragments are enclosed in quotation marks, that's okay—the tag, the "she said," will serve as the subject and the verb of the sentence.
5. The tricky area to punctuate is located right before and after the tag. Statements that would usually be followed with a period are joined to the tag with a comma. Questions and exclamations use those marks as usual, but never are two different marks of punctuation used to join the dialogue to the tag, like a question mark and a comma. Sometimes tags interrupt one sentence, so commas are used on either side. Sometimes tags end one sentence, and another sentence begins right after it. Can you find examples of both in the example above? What marks are used?
6. A pair of quotation marks will continue to work for as long as the person speaks, for as many sentences as the person speaks. Every single sentence is not enclosed in separate quotation marks.
7. At the end of a quoted passage, commas and periods always go inside the quotes.
8. Indirect quotes, which simply report what was said but are not the speaker's exact words, do not use quotation marks: Tim said that he was planning a field trip.

Using dialogue within your narratives adds interest, characterization, and realism to your work. (Try it! You'll like it! Or at least your readers will.)

- Quotes from other sources

Introduction to quote>	**In his book *Night*, Elie Wiesel speaks of his experience in the concentration camp:**
Quote from book>	**"Never shall I forget those moments that murdered my God and my soul and turned my dreams to ashes. Never shall I forget those things, even were I condemned to live as long as God Himself. Never."**

(Notice the colon, which is used to introduce the quoted passage.) This quote was not originally within quotes in the book.

To quote things that are already quoted, you simply turn the original quotes to single quotes and put double quotes around the whole thing.

Introduction to quote> Quote that used> quotes already	**Elie Weisel remembers how he responded when he was once asked why he prayed: "Why did I pray? Strange question. Why did I live? Why did I breathe? 'I don't know,' I told him, even more troubled and ill at ease. 'I don't know.'"**

Single quotes are typed using the apostrophe key.

- Quotations used for "special" words

Quotes may be used to denote words used in a special or ironic sense, but use these sparingly.

Example: His idea of a "vacation" was a weekend spent in a rundown fish camp motel.

- Quotations used for titles

The title of a minor work is put in quotations when the title is written out in a sentence. *All titles are just capitalized correctly when they are on the cover of a report, book, essay, etc., actually acting as the title of the work.* Minor titles are of works that are not published by themselves; they are published with other minor works like themselves under the title of a major work. These are minor works: short stories, essays, poems, reports, articles in newspapers or magazines, songs, and chapters in a book. (Think about it: all of these things are published with others like them, right? They are minor titles.)

Example: I had to read the poem "My Last Duchess" and the seventh chapter in my English book, a chapter called "Quotations and Italics."

Italics

- Italics, that squiggly, cursive-looking writing, are used to indicate the titles of major works when those titles are written out in sentences. Major works are published independently and often contain minor works within them. These are major works: books, newspapers, magazines, plays/movies, albums/CDs. (Think about it: all of these contain minor works within them, chapters, articles, acts or scenes, songs, etc., right?)

Example: I had started to read an article entitled "What's Your Sex I.Q.?" in *Cosmopolitan* magazine, but I had to stop to finish reading the last act of *Hamlet* for homework.

Note: In handwriting, italics are indicated by underlining. When typing, either italics or underlining may be used.

- Italics are also used to indicate names of ships, airplanes, and spacecraft, and the titles of works of art.

Example: The Smithsonian museum has pieces of the *Titanic* and the space shuttle *Challenger*, and this month it has a visiting exhibit of Piccaso's work, including *The Old Guitarist.*

- Occasionally, italics may be used to show special emphasis. (I used them in the last note concerning the use of quotations. What did I want to emphasize?)

Example: What *are* you doing?" my mother asked as she caught me painting my chest green.

- Words of a foreign language and words that are referred to as words are italicized.

Example: We cooked some *katsudon* and counted the number of times we said the word *good* as we ate it.

CHAPTER 7: EXERCISE 1

Choose a partner from someone in your class, preferably someone that you don't know well. The two of you are to sit next to one another, pass one sheet of paper back and forth, and write a conversation between the two of you. If your name is Mark, you'll use tags like "Mark said" to note which lines you wrote. Anything you can say out loud you can write as dialogue, even "I can't think of anything to say." Check one another's use of indention and punctuation as you go. Vary where you put your tags and use description to add detail as to how it was said and what was going on. Both people should write at least eight responses that are each a few sentences long.

CHAPTER 7: EXERCISE 2

Read all of the following conversation first. Then add punctuation where necessary to the conversation between two students. Put arrows before the lines that need to be indented.

Would you like to go to the movies with me   Tony asked

I don't know. What are you going to see   replied Mallory

Shrek the Third   responded Tony

Yeah, I'd love to go with you   said Mallory   because I've heard that it is really funny.

What time will you be by to pick me up

About 7:00   said Tony   so we'll have time to go by Wendy's and eat a bite too.

That sounds great   exclaimed Mallory   I'll be ready.   What are you wearing

Just some jeans and my Andrew College sweatshirt replied Tony Don't get all dressed up

Okay responded Mallory but can I bring someone with me

Who Tony inquired

Just my roommate said Mallory She never gets asked to go anywhere

That isn't what I had in mind said Tony

Aww, come on Mallory pleaded She's a lot of fun

Then why doesn't she get asked to go anywhere Tony replied Besides, I had plans for us after the movie

Oh? Like what Mallory asked

Some private time just between you and me said Tony with a gleam in his eye

Well, I guess she'll just have to stay home then Mallory responded slyly

CHAPTER 7: EXERCISE 3

Your instructor will read aloud this conversation between two students. Listen carefully and on a separate sheet of paper, write it out, indenting and punctuating as needed. You may check your work against this when you finish (if your instructor doesn't collect it for a grade.) Instructor, you may want to read the whole conversation aloud first and then read each line separately as students write it out.

"I'm going home this weekend. Are you?" asked Tom.

"No, I have a game Saturday afternoon," replied Jill. "Where do you live? Do you have to go far?"

"I live in Macon," answered Tom. "It's not that far. I can be home in an hour and a half."

"That's good," said Jill. "What will you do this weekend?"

"I'll just hang out with my friends and catch up on my sleep," Tom replied. "It's too bad you have to stay here."

"It will be fun," Jill responded. "The softball team is going into Albany Saturday night to the movies. We will have a great time."

CHAPTER 7: EXERCISE 4

Put each title in quotations or italics (since you're handwriting, that means underline) as necessary.

My summer reading list was ridiculously long. I had to read several short stories: Young Goodman Brown, Everyday Use, A Rose for Emily, and The Crysanthemums. Then there were many poems: I Hear America Singing, Bereft, To an Athlete Dying Young, The Red Wheelbarrow, and many others that I can't remember. Oh, then I had to catch up on magazine titles: The Southern Review, Writer's Pen, and The Writer's Craft. On top of that, I had to read several articles my instructor chose, including The Overuse of Commas, Much Ado About Note-taking, and So, You Want to Be a Writer. Finally, I had to read two novels and compare their styles: The Book of Ruth and Songs in Ordinary Time. I don't want to read anything else anytime soon. I'm going to catch up on some movies that I've been wanting to see, like Norbit, Night at the Museum, Daddy's Little Girls, and The Hitcher. Yeah, and I want to listen to some new CDs: Minutes to Midnight, by Linkin Park; It Won't Be Soon Before Long, by Maroon 5; and Double Up, by R. Kelly. I'm ready to relax!

CHAPTER 7: EXERCISE 5

Bring a magazine to class. Any kind will do, *Sports Illustrated, Cosmopolitan*, etc., so long as someone is quoted, speaking, inside. Study the earlier examples given in the clarifications of **Quotes from Other Sources.** Find a passage in your magazine that *does not already use quotes*; then write an introduction to the quote, put a colon, and supply the quote, using quotation marks correctly. Next, find a passage in your magazine that *does already use quotes*; then write an introduction to the quote, put a colon, and supply the quote, using single quotes and double quotation marks correctly. Show your instructor the originals and your versions.

Name ____ Date____________

CHAPTER 7: PROGRESS CHECK QUIZ

Read all of the following conversation first. Then add punctuation where necessary to the conversation between two people. Make your marks clear and distinct. Put arrows before the lines that need to be indented.

Where do you want to go to eat asked Simon as he drove the car away from the curb

I'm easy. It doesn't matter to me

Oh, I don't know said Jessica as she checked her makeup in the visor mirror Why don't we try that new Mexican restaurant on State Street

I heard that place was nasty Simon responded The guy that put in the walk-in refrigerator told me that it was really dirty. Let's go anywhere but there

Okay, how about the Jamaican restaurant Jessica asked You love their jerked chicken

I just ate there for lunch today Simon replied Hey, I know. Let's go to Andre's. You love the chicken picata, and we haven't been there lately

I had Italian last night, so I really don't want it again Jessica sighed Why is this so difficult

Simon and Jessica rode in silence for a couple of minutes. Then Simon looked over at Jessica as he turned into the parking lot of the local grocery store.

We'll have a steak and baked potato that I'll cook at home Simon said with finality.

Sounds good Jessica said, relieved.

Do the following types of titles require quotation marks or italics if they are written out in a sentence? Put Q or I.

Magazine title ______________
Play title ______________
Poem title ______________
Work of art______________

Newspaper article ______________
Movie title ______________
Short story title ______________
Magazine article ______________

CHAPTER *eight*

Apostrophes and Hyphens

JUSTIFICATION: WHY DO I NEED TO KNOW THIS?

Apostrophes and hyphens are used quite often in college writing (and in day-to-day writing, actually), and a competent writer must know how to use them.

CLARIFICATIONS

Apostrophe

Apostrophes are used to show possession.

- If the original word does not already end in s, add 's.
  Dan's car   one student's book   the children's toys   the men's club

- If the original already ends in *s*, add just the apostrophe.
  Frances' hair   two students' books   the Joneses' house   two kids' toys

* Many authorities put 's on all proper nouns (generally people's names), no matter how they end. You may do either (for people's names) but be consistent: Chris's girlfriend or Jesus' name

- For compounds, if both own something together, make only the second one possessive.
  Tim and Julie's house (one house owned by both)

- For compounds, if each owns something separately, make both possessive.
  Fred's and Mark's cars (two cars, each owns one)

- Personal pronouns do not use apostrophes; they already show possession.
  his   hers   yours   theirs   its*

Apostrophes are used to show omissions in contractions and numbers. Put the apostrophe where the omission occurs.

wouldn't	could've	she'll	he'd
class of '08	y'all	goin'	it's*

*Remember, *it's* is *it is*. *Its* shows the possessive form of *it*.

Apostrophes are also used to show certain plurals.

How many *no*'s did you utter today?
The old typewriter would not type e's or o's.

Hyphens

Hyphens are used to join two or more words serving as a single adjective before a noun.

chocolate-filled cookies (not chocolate cookies, but chocolate-filled cookies)
eighteen-year-old student
face-to-face class

Hyphens are used in writing compound numbers from twenty-one to ninety-nine. (Use this when writing out the amount on checks and documents.)

Eighty-six and 00/100 dollars   twenty-seven   forty-four

CHAPTER 8: EXERCISE 1

Write out the possessive forms of the following. Be careful—do you put the apostrophe before or after the *s*?

Ex.: letter of Lincoln   Lincoln's letter

1. knee of Mike ______________
2. sale at Rite-Aid ______________
3. murder of Tabitha ______________
4. bike of James ______________
5. tires of the car ______________
6. guns of the agents ______________
7. pride of the Americans ______________
8. vacation of the Roberts ______________
9. dogs of Sandy ______________
10. mother of Zak Phillips ______________
11. review of Norton ______________

12. restroom of the women ______________________

13. fans of the Minnesota Twins ______________________

14. boat of David ______________________

15. house of Lowell and Kelly ______________________

16. map of Allen ______________________

17. paper of Kate ______________________

18. tour bus of the Beastie Boys ______________________

19. car of Mary and Marc ______________________

20. shouts of the freshmen ______________________

CHAPTER 8: EXERCISE 2

Add any needed apostrophes and circle them. If none are needed, put **C** next to the number of the sentence. Be careful—everything that ends in *s* does not need to be made possessive.

1. How many *of* s did you use in that sentence?
2. Sadie and Nates garden needs watering.
3. I couldnt come up with better answers, so I didnt respond.
4. John asked for his autograph but got hers instead.
5. Terra, its too late to give the bird its medicine.
6. When do yall want yalls pecan pie?
7. Have you used * s in a paper?
8. Mens Wearhouse had its suits on sale, and he shouldve bought one.
9. She got Babs card, and shed have called her if shed had Babs number.
10. John Edwards campaign doesnt appear to be winning over any Georgians.
11. The professors leather briefcase had lost its shine and one of its buckles.
12. Many of the students brought their books and sat on the benches outside to study.
13. Most of the dogs hair had fallen out, and its eyes were matted and lifeless.
14. I shouldve gone when I couldve, and then I wouldnt be in this mess.

15. Jans and Missys computers were left on, and the lightening zapped their hard-drives.
16. The Star Teachers luncheon was held at the college, and every teacher was given a plaque.
17. Were you in the Class of 08 or 09?
18. All of the librarys books need their covers dusted, and the books that are on these tables need to have their bindings repaired.
19. Georges new job required that he wash all the dogs bowls, clean all the dogs pens, and exercise the dogs outside everyday.
20. My research papers Works Cited page is all wrong; its the spacing that I have to work on over the weekend.

CHAPTER 8: EXERCISE 3

Add any needed apostrophes and hyphens; then circle them.

1. Vivian remembered her two year old sons ice cream covered face.
2. Marks Melon Patch sells home grown vegetables and fruits.
3. Many kids life or death struggles could be eased with guaranteed health insurance coverage.
4. Kristie was grief stricken at the news of her grandmothers death.
5. Many characters in Jareds novels are larger than life heroes who seem to have few faults.
6. All womens, mens, and childrens clothes and shoes are on sale this week at Sears.
7. With the mail in offer, I could get the Shrek keychain.
8. Alans moms cake won first prize in the carnivals baking contest.
9. Felicity took her doctors advice and began eating a high fiber diet to help her constipation.
10. Dad, I need eighty six dollars to pay off my drum set, but Ill give you forty two dollars when I get paid next week.

11. The top of the line model was thousands of dollars more than the mid grade model.
12. Susans allergy symptoms were eased with her doctors recommendations.
13. Herschels drug free weight loss program was a huge success on the Internet.
14. The product offered a money back guarantee to its customers.
15. Sarah and Mitchs kids are smarter than theirs.
16. The Class of 80 will have its reunion at Sherri and Pats house.
17. We need to install high density foam insulation to the houses exterior to improve its energy efficiency.
18. All of the residents suitcases must be inspected for weapons before opening the buildings doors.
19. All of the F s on that teachers set of report cards had to be compared to her grade book to ensure accuracy.
20. Julie was glad to have the hand me down clothes from Sabrinas closet; she planned to sell them to the second hand clothes store downtown.

Name ______________________________ Date__________

CHAPTER 8: PROGRESS CHECK QUIZ

Write out the possessive forms of the following. Be careful—do you put the apostrophe before or after the *s*?

Ex.: letter of Lincoln — Lincoln's letter

1. tests of the two students ____________
2. dog of Sandra and Scott ____________
3. faces of the audience ____________
4. flowers of the four bridesmaids ____________
5. grades of Sonja ____________
6. clasp of the bracelet ____________
7. knives of the two cooks ____________
8. bathroom of the women ____________

Add any needed apostrophes and hyphens; then circle them.

9. Wayne was given twenty four hours to find sixty two volunteers.
10. The product had a money back guarantee for its customers satisfaction.
11. Mikes lab coat was found in Kathys cars trunk.
12. The models full bodied hair was difficult to insert between the plates of the hair straightener.
13. We shouldve used Terrys and Shedricks trucks to haul the dirt.
14. After the finale, the three judges decision was published in the peer reviewed journal.
15. For the party, Jerome preferred to barbeque his own country style pork ribs instead of his brothers all beef hamburgers.

CHAPTER *nine*

Capitalization

JUSTIFICATION: WHY DO I NEED TO KNOW THIS?

Competent writers must know how to use capitalization correctly. Otherwise, they will appear poorly educated to those who know better, usually people whom the writers would like to impress favorably, people like professors, employers, coworkers, and possibly even prospective dates. Capitalize correctly; it reflects your level of education.

CLARIFICATIONS

Capitalize proper nouns.

1. Names of specific people, places, and things

people:	Rosie O' Donnell	Donald Trump	Prince
places:	Andrew College	the Alamo	Elm Street
things:	Jeep Cherokee	Rolaids	Pepsi

2. Geographical place names

Antarctica	Arctic Circle	Lake Eufaula	the South*
the Middle East	Pacific Ocean	Equator	Cuthbert

*Directions are not capitalized: We drove east for two hours.
Specific place: The Far East has always fascinated me.

3. Countries, languages, peoples, and proper adjectives derived from them

England	English*	Asia	Asian
Russia	Russian	Hispanic	Latinos

*The names of general school subjects are not capitalized (unless they are proper adjectives), but the specific names of particular courses are.
Ex.: I'm taking math, tennis, English, and Biology 121.

4. Religions and their derivatives, names (and pronouns) for the Supreme Being

Islam	Islamic	Baptist	Catholic
Jewish	God	Allah	Jesus and His followers

5. Specific groups
   Friends of the Library   Phi Theta Kappa   House of Representatives
   United States Marine Corps   Hell's Angels   Veterans of Foreign Wars

6. Historical events, documents, periods
   World War II   Declaration of Independence   Iron Age

7. Days of the week, months, holidays
   Monday   July   Flag Day   Spring Break*
   *Seasons are not capitalized: summer, winter, fall, spring.

8. Titles for people
   President Jimmy Carter   Dean Chip Reese   Reverend Billy Graham

9. Words used in place of a family member's name but not those used to denote relationships between family members
   I told Dad that I needed more money.
   I told my dad that I needed more money.
   I don't want Uncle Bob to sleep in my room.
   I don't want my uncle to sleep in my room.

10. Titles—Capitalize the first and last word of any document title and all other words that are not articles (a, an, the), prepositions, conjunctions (and, but, or, nor, for, yet, so), or *to*.
    Footwear in the Suburbs of America: More than Just a Pair of Shoes
    Cosmo Girls: American Young Women and Their Sexuality
    Russia: The Land of the Czars
    (title) : (subtitle)*
    *Capitalize the first word of a subtitle, too.

11. Capitalize the first word of any sentence (duh).

CHAPTER 9: EXERCISE 1

Add any needed capitalization; just write over the lowercase letter that is there. If no capitalization is needed, put a **C** next to the number of that sentence.

1. My uncle is the pastor at the local methodist church in our hometown.
2. My uncle, reverend thomas willis, is the pastor of smithville united methodist church in smithville, florida.
3. I like to watch that female chef that is on that cooking show on channel 62.
4. I like to watch giada de laurentis during her show *everyday italian* on the food network.
5. If you take oak street north for five miles, you should find a starbucks in the mountain view mall.

6. The world's image of the american soldier was damaged severely by the photos of iraqi prisoners being abused at the abu ghraib prison.
7. The university of north carolina's women's la cross team played the university of virginia's team on saturday, june 16.
8. On *the andy griffith show,* aunt bea served as a mother-figure to both opie and andy, kind of an interesting freudian idea.
9. The science fiction classic *star trek* encouraged universal harmony; captain james t. kirk, an earthling, was best friends with mr. spock, a vulcan.
10. Abdullah prayed that allah would protect josh during his trip to iran and that he would shower his blessings on josh's host family.
11. During the great depression, many americans looked to president roosevelt and his new deal program to help them survive.
12. My roommate, suki, is from kyoto, japan, and she is teaching me to cook japanese dishes.
13. I saw the new video from fall out boy on mtv; the apes were really funny.
14. Does anybody know what richard petty really looks like? The nascar star always wears dark glasses and a hat, an ingenious disguise that allows him to be a regular guy if he takes them off.
15. The space shuttle *atlantis* is heading back to earth after a trip to the international space station.
16. Sue won a week-long trip to aruba on the carnival cruise ship named *fantasy.*
17. Horrible flooding occurred in northern texas, and several people lost their lives in gainesville, according to the weather channel report.
18. Paul called his graduate thesis "an ethnographic study of the yanomami tribe: a people in transition from an ancient to a modern society."
19. When prince william and prince harry were interviewed by matt lauer for the today show, the british press were jealous; the royals didn't give interviews to them.
20. A massive u.s. offensive began after the bombing of a shiite mosque in iraq on friday.

CHAPTER 9: EXERCISE 2

Add any needed capitalization; just write over the lowercase letter that is there. If no capitalization is needed, put a **C** next to the number of that sentence.

1. My husband, david, and i took my dad fishing on father's day this year.
2. My father is 78 years old and has always been a physical fitness fanatic.
3. A retired marine, pop used to be a boxer for the u.s. marine corps.
4. He boxed at madison square garden, was a golden gloves champion, and is proud to say that he was never knocked out.
5. Still active and fit, pop walks a mile and a half every day around his neighborhood: down jenkins road, up james drive, and then around county line road.
6. The fishing trip to lake eufaula, part of the body of water that marks the state line between alabama and georgia, was a nice change for my dad.
7. On father's day morning, we took david's eighteen-foot triton boat with mercury outboard motor and headed west.
8. Needing some bait, we stopped at top bait, a fishing store in cuthbert, georgia.
9. Hopping out of the ford f-150 truck, pop went into the store to check out the merchandise.
10. He had a good time looking at all of the gadgets, especially the rapala lures, plano tackle boxes, and columbia sportswear.
11. The store owner, charles, asked us where we picked up "rambo."
12. I hadn't thought twice about pop wearing his u.s. marine corps desert storm hat.
13. We all laughed; my father did look like a geriatric rambo in his tough-guy hat.
14. When david told charles that pop was a former marine corps boxing champion, charles was quite impressed, and my dad grinned with pride.

15. Arriving at bagby state park, the boat launch that is maintained by the u.s. army corps of engineers, my father hopped out of the truck to help guide the boat into the lake.
16. We fished for five hours and didn't catch a thing, but we had a great time laughing about it.
17. We tried deep water and shallow water; we went down to drag nasty, pataula creek, and the edges of the u.s. intercoastal waterway.
18. Pop did hook a catfish that was only four inches long; i can't repeat the obscene comment my father made about that.
19. Since the boat was loaded with cokes, sandwiches, diet cokes, little debbie snacks, corona beer, and chips of all sorts, no one went hungry.
20. My husband and my father both had a great father's day outing, and i have another precious memory of my dad and me.

CHAPTER 9: EXERCISE 3

Add any needed capitalization; just write over the lowercase letter that is there. If no capitalization is needed, put a C next to the number of that sentence.

1. In charleston, south carolina, nine firemen lost their lives after heroically saving two men from a fire at the sofa super store.
2. Fashion model iman is married to rock star david bowie.
3. cole bought a dell computer online and bought software at circuit city.
4. Another model, eva mendes, is a beautiful hispanic woman who is also breaking into acting.
5. Music legend stevie wonder has won twenty grammy awards, and, after a ten-year hiatus, he has a new release called *a time to love.*
6. The newest controversial film from michael moore is called *sicko*, and it may make health insurance companies, like aetna and humana, really upset.
7. When my aunt came to visit, i got some chicken from kentucky fried chicken, made some german potato salad, and an italian cream cheese cake.

8. In the term paper entitled "the torment of bulimia: a personal account," the author discussed her own battle with the common eating disorder.
9. For supper, tom made hillshire farms sausage, bush's baked beans, and a caesar salad.
10. In its charter, the sorority delta kappa zeta specified that the college would benefit from its service projects.
11. A woman from idaho named susan hobbs promoted a law that allows young mothers to safely abandon unwanted babies at their local hospital or firehouse.
12. When judge karl ross sentenced the rapist to twenty-five years in prison, the judge didn't want him to have any possibility of parole.
13. On fox news last night, paris hilton was shown going to jail without an armani suit or gucci handbag.
14. tony snow, the white house press secretary, discussed president bush's veto of the bill.
15. Where are you going for the labor day holiday this fall, panama city or atlanta?
16. Before cousin milton left for new york last summer, we ordered a sicilian pizza from guido's pizzeria and had a little party.
17. In my class called servant leadership 212, we read william b. turner's book entitled *the learning of love: a journey toward servant leadership.*
18. I took mom to the doctor, picked up my dry cleaning from dan's cleaners, went grocery shopping at publix, and still had time to go to custom cuts for my hair appointment.
19. steven was named star student and chose his long-time friend and mentor, his football coach, don beard, as his star teacher.
20. When you first saw *the wizard of oz,* were you afraid of the flying monkeys?

Name ______________________________ Date__________

CHAPTER 9: PROGRESS CHECK QUIZ

Add any needed capitalization; just write over the lower-case letter that is there. If no capitalization is needed, put a **C** next to the number of that sentence.

1. In the summer of 2004, i went to lithuania, latvia, and russia on a fulbright scholarship from the u.s. department of education.
2. I had never flown on an airplane before, so the flight across the atlantic on the boeing 747 was pretty exciting for me.
3. we attended seminars hosted by the university of latvia and the other universities we visited; one very interesting lecture was entitled a brief historical and some contemporary aspects of latvia, by eriks leitis.
4. we were there during june for yannis, which is a celebration of the longest day of the year, and we were there during white nights, which is a period when the sun never really sets; local citizens party all night.
5. we went to st. petersburg, russia, where we visited peterhof, the summer home of peter the great, and we went to moscow, where we toured the kremlin.
6. the kremlin contains the modern buildings of the russian government, including the offices of vladimir putin, and the cathedrals where all of the czars were crowned, including the last of the romanovs.
7. while in moscow, we toured red square, which includes lenin's tomb and st. basil's cathedral, the one with the beautifully-colored onion domes.
8. an unforgettable aspect of my trip was being left by the tour bus outside of the novodevichy monastery; I walked across moscow alone, finding my way back to the hotel russea, which, thankfully, was right next to the kremlin.
9. through lectures and slide shows, i have shared with others what i have learned, and i am applying to participate in a trip to china and singapore for the summer of 2008.